Contents

1

Introduction to Biotechnology

Biotechnology is the applications of scientific and engineering principles to the processing of material by biological agents to provide goods and services. Biotechnology can be viewed as a group of useful, enabling technologies with wide and diverse applications in the industry, commerce and environment. Historically, biotechnology evolved as an artisanal skill rather than a science, exemplified in the manufacture of beers, wines, cheese etc. Today biotechnology is applied to manufacturing processes used in health care, food and agriculture, industrial processes and environmental cleanup, among other applications.

Old and New Biotechnology

The origin of biotechnology goes back to 6000 B.C with the story of the use of biological systems for the fulfilment of human needs, when Sumerians and Babylonians fermented a kind of beer. Beginning with fermentation of beer, the use of biological processes then experienced has undergone many changes over the centuries. But greatest revolution has taken place in the 1970's and 1980's, when a product of interaction between the science of biology and technology came into wider existence. This relationship thus got the name Biotechnology. The origin of biotechnology can be traced back to prehistoric times when microorganisms were already used for processes like fermentation, formation of yoghurt and cheese from milk, vinegar from molasses, production of butanol and acetone from starch by Clostridium acetobutylicum.

Fast forward to modern biotechnology in the last few hundred years: Some of the most important modern biotechnologies were the development of vaccines and antibiotics such as the production of antibiotics like penicillin from Penicillium notatum. Jenner, 1795, is credited with invention of vaccination in Western medicine. Jenner found that vaccination with cowpox virus makes people immune to smallpox, which is caused by smallpox virus (though of course the germ theory of disease and the concept of viruses did not come until much later). Sometime between 1795 and 1930, the cowpox virus used for vaccination against smallpox was replaced with vaccinia virus, a contaminant of unknown origin that is not the same as cowpox virus. The terms vaccinia and vaccination are based on the Latin name for cow (vacca).

In the 1960s and 70s, our knowledge of cell and molecular biology reached the point where we could begin to manipulate organisms at those levels. Essentially, we learned how to move genes around, at will. Manipulating organisms to our advantage is not new. What is new is how we are manipulating them. The key to new technology was recombinant DNA, the term biotechnology implied the use of recombinant DNA. But the meaning of the word has become broader by the use of cells and biological molecules to solve problems or make useful products.

Nowadays it can mean almost anything that has to do with biology and technology. Defenders of the use of genetically engineered plants and animals are fond of the point that biotechnology is not new and of the idea that manipulating organisms to our advantage (by selective breeding) is not new. We will focus on other aspects of biotechnology i.e. human health, which experts estimate is the aim of more than 80% of biotech research funds (but that is difficult to measure). Biotechnology makes its contribution in all fields related to health care, especially in the fields of Microbiology, Genetics, Biochemistry and Immunology which have principal contribution to the medical field for many years. Biotechnology research is aimed towards:

— Diseases caused by infective agents, including viruses

— Diseases due to imbalance in body's natural chemistry

— Genetic disorders

— AIDS, Tumor and Cancer Therapy

— Organ Transplantation

The numerous tools and sciences that are the basis for biotechnology are kind of overwhelming, but remember no one person knows everything about all of this which are also fundamental to success in biotechnology. Through the centuries with the increase in population of the world and consequently increase in the demand for food supply, various novel methods have been used to increase the production of food and also improve the quality and there by fulfil the demands of the increasing population. Since plants are the key to life on earth, they directly supply 90% of human calorie intake, and 80% of the protein intake and the remainder being derived from animal products, although these animals have also derive their nutrition from plants.

Of the three thousand plant species which have been used as food by man, the world now depends mainly on around twenty crop species for the majority of its calories, with 50% being contributed by eight species of cereals. Minerals and vitamins are supplied by a further thirty species of fruits and vegetables. Most important of the staple foods are the cereals, particularly wheat and rice, with more than one-third of all cultivated land used to produce these two crops.

As the population continues to expand, there has been a concern over the large number of people that the world agriculture can support. It has been calculated that the earth can support about 15 billion people on a strictly vegetarian diet, or five billion on a mixed diet. The population is expected to be fifteen billion by 2025. The farming practices and crops cultivated today have developed over a relatively short time span. Crop plants of today have changed in a number of ways so that they now bear very little resemblance to their wild type ancestors. These changes have come about through selection, either conscious or unconscious, for traits which are advantageous to the people growing the crops. Thus modern wheat do not disperse their seeds or legumes do not have pods which burst open.

Today, varieties are the result of generation of plants cultivated under ideal conditions from the man's point of view. From the beginning of crop cultivation to the late nineteenth century, all improvements in the species used were brought about by those who were directly involved, i.e. farmers themselves. In the following 100 years, the laws of genetic inheritance and rules governing species variation were laid down by Mendel, Darwin and others which redefined the breeding techniques by making them predictable and therefore quicker, more precise and more productive. Furthermore, despite the implementation of breeding techniques, the time taken to produce and test varieties is an important limiting consideration.

Similar to the improvements in plant production, there has been a considerable advances in animal production, such as breeding animals for disease resistance, high productivity of milk in bovine animals, etc. Microorganisms have been extensively used in various industries for the production of vaccines for various diseases, production of antibiotics, in fermentation processes etc. Biotechnology thus consists of a variety of techniques, designed to genetically improve and/or exploit living systems or their components for the benefit of man. Infact, biotechnology is the product of interaction between sciences of biology and technology.

However, biotechnology got a boost in the 1970's with the discovery of restriction enzymes which led to the development of a variety of gene technologies and is thus considered to be the greatest scientific revolution of this century. The prokaryotes, eukaryotic algae, glycophytes and halophytes all are likely to contribute in commercially viable future propositions. Therefore Biotechnology is the application of biological organisms, systems or processes to manufacture and service industries, comprises a number of technologies based upon increasing understanding of biology at the cellular and molecular level.

The following are the Areas of applied biotechnology: In 1885, a scientist named Roux demonstrated embryonic chick cells could be kept alive outside an animal's body. For the next hundred years, advances in cell tissue culture have provided fascinating glimpses into many different areas such as biological clocks and cancer therapy. Monoclonal antibodies are new tools to detect and localise specific biological molecules.

In principle, monoclonal antibodies can be made against any macromolecule and used to locate, purify or even potentially destroy a molecule as for example with anticancer drugs. Molecular biology is useful in many fields. DNA technology is utilised in solving crimes. It also allows searchers to produce banks of DNA, RNA and proteins, while mapping the human genome. Tracers are used to synthesise specific DNA or RNA probes, essential to localising sequences involved in genetic disorders.

With genetic engineering, new proteins are synthesised. They can be introduced into plants or animal genomes, producing a new type of disease resistant plants, capable of living in inhospitable environments (i.e. temperature and water extremes,). When introduced into bacteria, these proteins have also produced new antibiotics and useful drugs. Techniques of cloning generate large quantities of pure human proteins, which are used to treat diseases like diabetes. In the future, a resource bank for rare human proteins or other molecules is a possibility.

For instance, DNA sequences which are modified to correct a mutation, to increase the production of a specific protein or to produce a new type of protein can be stored. This technique will be probably play a key role in gene therapy. Over the past few years a number of methodologies have come that would seem to have much more to offer in terms of advancing current research in the plant sciences, and exploiting the knowledge gained to develop new crops. The first of these areas is concerned with manipulation and subsequent growth of cells, tissues, organs and naked plant cells (protoplasts) in tissue culture.

The second field is genetic engineering or recombinant DNA (rDNA) technology which has grown out very fast from the work initially carried out on microorganisms. With genetic engineering, scientists have more exact methods for breeding better livestock and crop varieties. This technology allows for the detailed manipulation of genes. These two areas of research have in recent years become associated with the general field of biotechnology and are potentially applicable to a wide variety of plant species, as well as offer a precision in manipulating genetic material.

Recombinant DNA techniques in particular have already contributed much to the elucidation of basic mechanisms in plants at the molecular level. The term genetic engineering refers to a number of new techniques involving the transfer of specific genetic information from one organism to another. These techniques do not rely on sexual methods, but instead involve genetic manipulation at the cellular and molecular levels. These are the non-sexual methods for gene transfer.

The technique includes recombinant DNA manipulations, monoclonal antibody preparation, tissue culture, protoplast fusion, protein engineering, immobilised enzyme, cell catalysis, sensing with the aid of biological molecules, etc. Biotechnology is still in the process of early development. Only in last fifteen years progress have been made by microbiologists and genetic engineers, and we are hopeful to solve many fold problems of the present day, specially energy and food crisis to cater the need of growing population of the world.

MODERN FACTS OF BIOTECHNOLOGY

Now we will be looking at the various facts about biotechnology, the perception of public and its global impact. Biotechnology is said to be the third wave in biological sciences and represents such an interface of basic and applied sciences, where gradual and subtle transformation of science into technology can be witnessed. Consequently, the new term "biotechnology" has emerged which includes processes that exploit versatile metabolic machinery or components of living organisms to produce valuable metabolites from renewable resources. This is very timely since due to the increasing human population, the supply of many valuable raw materials is diminishing.

The new technology can be used to create novel resources and materials. Biological systems can now be exploited at many new frontiers. The objective of studying the modern facts is that we will be able to know the different fields in which the biotechnology works and what are the perceptions of the public regarding the biotech products. There are various modern facts regarding biotechnology which are as follows: Over 200 million people worldwide have been helped by more than 80 biotechnology drug products and vaccines approved by the U.S. Food and Drug Administration (FDA).

— There are more than 350 biotechnology drug products and vaccines currently in human clinical trials and hundreds more in early development in the United States. These medicines are designed to treat various cancers, Alzheimer's, heart disease, multiple sclerosis, AIDS, obesity and other conditions.

— Biotechnology is responsible for hundreds of medical diagnostic tests that keep the blood supply safe from the AIDS virus and detect other conditions early enough to be successfully treated. Home pregnancy tests are also biotechnology diagnostic products.

— Consumers are already enjoying biotechnology foods such as vine-ripened, longer-lasting tomatoes and better-tasting carrots and peppers. Hundreds of biopesticides and other agricultural products are also being used to improve our food supply and to reduce our dependence on conventional chemical pesticides.

— Environmental biotechnology products make it possible to more efficiently clean up hazardous waste without the use of caustic chemicals.

— Industrial biotechnology applications have led to cleaner processes with lower production of wastes and lower energy consumption, in such industrial sectors as chemicals, pulp and paper, textiles, food and fuels, metals and minerals and energy. For example, much of the denim produced in the United States is finished using biotechnology enzymes.

— DNA fingerprinting, a biotech process, has dramatically improved criminal investigation and forensic medicine, as well as afforded significant advances in anthropology and wildlife management.

— Market capitalisation, the amount of money invested in the U.S. biotechnology industry, increased 4 percent in 1998, from $93 billion to $97 billion.

— Approximately one-third of biotech companies employ fewer than 50 employees. More than two-thirds employ fewer than 135 people.

— The U.S. biotechnology industry currently employs more than 153,000 people in high-wage, high-value jobs.

— Biotechnology is one of the most research-intensive industries in the world. The U.S. biotech industry spent $9.9 billion in research and development in 1998. The top five biotech companies spent an average of $121,400 per employee on R&D. This compares with an average of $31,200 per employee for the top pharmaceutical companies.

— The biotech industry is regulated by the Food and Drug Administration (FDA), the Environmental Protection Agency (EPA) and the Department of Agriculture (USDA).

Public Perception of Biotechnology

Public perception means that what the people world-wide

think about the bio-products evolved from new technologies. It is clear that biotechnology has its roots in the past and has large, highly profitable, modern industrial outlets of great value to society, e.g. The fermentation, biopharmaceutical and food industries and now the question arises how far are the new technology products beneficial to the mankind. The main reasons must have been associated with the rapid advances in molecular biology, in particular recombinant DNA (rDNA) technology, which is now giving bioscientists a remarkable understanding and control over biological processes.

As you very well know that Genetic engineering is viewed as a branch of modern science which will have profound impacts on medicine, contributing to the diagnosis and cure of hereditary defects and serious diseases, the development of new biopharmaceutical drugs and vaccines for human and animal use, the modification of microorganisms, plants and farmed animals for improved and tailored food production and to increased opportunities for environmental remediation and protection.

Genetic Engineering

However, genetic engineering is surprisingly being subjected to a massive level of criticism from a deeply suspicious public. While the American public seem to have a greater acceptance of the potentials of genetic engineering, in Europe the technology appears to arouse deep unease among many consumers. Consumers demonstrate concern about 'unknown' health risks, possible deleterious effects on the environment and the 'unnaturalness' of transferring genes between unrelated species.

There is no doubt that many of the public or consumers are interested in the science of genetic engineering but are unable to understand the complexity of this subject. Furthermore, genetic engineering and its myriad of implications must not be beyond debate. Public attitudes to genetic engineering will influence its evolution and marketplace applications. It is important for public confidence for everyone to recognise (including scientists) that all science is fallible—especially complex biological sciences.

The following questions arise as far as the understanding of genetic engineering is concerned. What then must be done to advance public understanding of genetic engineering in the context of biotechnology? What does the public need to know and how can this be achieved to ensure that the many undoubted benefits that this technology can bring to mankind do riot suffer the same fate as the food irradiation debacle in the UK in the early 1990s?

While gamma-irradiation of foods was demonstrated to be a safe and efficient method to kill pathogenic bacteria, it was not accepted by the lay public following the Chernobyl disaster, since most were unable to differentiate between the process of irradiation and radioactivity. Effective communication about the benefits and risks of genetic engineering will depend on understanding the underlying concerns of the public together with any foreseeable technical risks. Over recent years there have been many efforts made to gauge the public awareness of biotechnology by questionnaires, Eurobarometers and Consensus Conferences.

Early studies highlighted public attitudes to the application of genetic manipulation to a wide range of scenarios. While medical applications were more generally acceptable others such as the manipulation of animal and human genomes were highly unacceptable. For centuries now man has been indirectly manipulating the genomes of plants and animals by guided matings primarily to enhance desired characteristics. In this way, food plants and animals bear little resemblance to their predecessors.

Safety Aspects of Genetically Modified Organisms

You must be aware that much debate is now taking place on the safety and ethical aspects of genetically modified organisms (GMOs) and their products destined for public consumption. The safety of the human food supply is of critical importance to most nations and all foods should be fit for consumption i.e. not injurious to health or contaminated. When foods or food ingredients are derived from GMOs they must be seen to be as safe as, or safer than, their traditional counterparts. The concept of substantial equivalence is widely applied in the determination of safety by comparison with analogous conventional food products together with intended use and exposure.

When such novel products are moving into the marketplace the consumer must be assured of their quality and safety. Thus there must be toxicological and nutritional guidance in the evolution of novel foods and ingredients to highlight any potential risks which can then be dealt with appropriately. Safety assessment of novel foods and food ingredients must satisfy the producer, the manufacturer, the legislator and the consumer.

The approach should be in line with accepted scientific considerations, the results of the safety assessment must be reproducible and acceptable to the responsible health authorities and the outcome must satisfy and convince the consumer! A comprehensive regulatory framework is now in place within the EU with the aim to protect human health and the environment from adverse activities involving GMOs. There are two Directives providing horizontal controls i.e.

1. contained use and
2. deliberate release of GMOs.

The contained use of GMOs means the use of the GMO is limited ie. the release into the environment of GMOs for experimental purposes (i.e. field trials) which is regulated under the Health and Safety at Work Act through the Genetically Modified Organisms (Contained Use) Regulations which are administered by the Health and Safety Executive (HSE) in the UK. Any deliberate release of GMOs into the environment i.e. the marketing of GMOs which is regulated in the UK by the Genetically Modified Organisms (Deliberate Release) Regulations, which are made under the Environmental Protection Act. Current examples could include the growing of GM food crop plants or the marketing of GM soybeans for food processing. In the UK the safety of all novel foods including genetically modified foods is assessed by the independent Advisory Committee on Novel Foods and Processes (ACNFP) which has largely followed the approach developed by the WHO and OEDC in assessing the safety of novel foods.

There is undoubtedly going to be a steady increase in the range of GM foods coming to the market in the US and in Europe, for example, Tomatoes with modified ripening; insect resistant potato, cotton, maize; glyphosphate tolerant soybean, oilseed rape and cotton; male sterile maize oilseed rape; high oleic acid soybeans etc. In all of the foregoing, the risk assessments of GMO products etc. have been made by experts and judged on the basis of safety to the consumer. However, it must be recognised that subject experts define risk in a narrow technical way, whereas the public or consumer without sufficient knowledge generally displays a wider, more complex view of risk that incorporates value-laden considerations such as unfamiliarity, catastrophic potential and controllability.

Furthermore, the public, in general, will almost always overestimate risks associated with technological hazards such as genetic engineering and underestimate risks associated with 'lifestyle' hazards such as driving cars, smoking, drinking, fatty foods etc. It is puzzling to note that food-related technologies tend to be perceived as high in risk relative to benefit when compared with other technologies. Perception of the risks inherent in genetic engineering may be moderated by recognition of the tangible benefits of specific products of genetic engineering that could be shown to have health or environmental benefits. Trust in the reliability of the information source is of major importance in any communication about risk and this is associated with perceptions of accuracy, responsibility and concern with public welfare. In contrast, distrust may be generated when it is assumed that the facts are distorted or the information misused or biased.

Impact of Biotechnology

In recent years revolution in biology has occurred due to the potential of biotechnology. Techniques have been developed to produce rare and medicinally valuable molecules, to change hereditary traits of plants and animals, to diagnose diseases to produce useful chemicals and, to clean up and restore the environment. In this way biotechnology has great impact in the fields of health, food/agriculture and environmental protection.

Due to rapid development the present situation is that there is no difference between pharmaceutical firms and biotechnology industry. However, approved products in the pipeline and renewed public confidence make it one of the most promising areas of economic growth in future. India offers a huge market for the products as well as cheap manufacturing base for export. Following are some of the areas where biotechnology has done the best.

1. *Health care*: The maximum benefits of biotechnology has been utilised by health care. Biotechnology derived proteins and polypeptides form the new class of potential drugs. For example, insulin was primarily extracted from slaughter animals. Since 1982, human insulin (Humulinl8) has been produced by microorganisms in fermenters. Similarly, hepatitis B vaccines viz., Recombivax HB@, Gunil8>, Shanvac@, etc. are the genetically engineered vaccines produced biotechnologically; since 1987, the number of biotechnology derived new protein drug has surpassed the new chemical drugs. Currently there are about 35 biotechnology derived therapeutics and vaccines approved by the USFDA alone for medical use, more than 500 drugs and vaccines to reach in market.

2. *Agriculture*: Biotechnology is making new ground in the food/agriculture area. Current public debate about BSTC, bovine somatotropin (a hormone administered to cows to increase milk production) typifies an example of biotechnology product testing public acceptance. Food biotechnology offers valuable and viable alternative to food problems, and a solution to nutritionally influenced diseases such as diabetes, hypertension, cancer, heart diseases, arthritis, etc. Biopesticides are coming to the market and their sales are increasing.

3. *Molecular Pharming*: It is a new concept where therapeutic drugs are produced in 11 animals, for example therapeutic proteins secreted in goat milk.

4. *Environment*: The natural biodegradability of pollutants present in environment has increased with the use of biotechnology. The bioremediation technologies have been found to be successful to combat with pollution problem.

INDUSTRIAL ORGANISMS

Chemical Technology and Novel Products

As a World War I necessity, German scientist Neuberg used largescale yeast fermentations to produce glycerol. On the British side, Weismann used Clostridium to produce acetone, butanol. Even though these fermentations were replaced by chemical technology, they provided the motivation for industrial production of citric acid by Aspergillus niger. The commercial production of penicillin in the 1940s by submerged fermentation was of great benefit to the Allied war effort and remains the technological triumph for producing microbial products of human interest.

Lets see how the living organisms are a source of many valuable products of diverse chemical structure and biological activity. For example, more than 6,000 antibiotics are produced by microorganisms and more than 150 have been commercialised. Antiparasitic agents, such as avermectins, are microbially produced. Microbial products like levastatin are used to lower blood pressure and cyclosporin, rapamycin and-FK-506 are stimulators of host defence against pathogens. Taxol is a plant product with antitumor activity.

Living Organisms and Foreign Proteins

It is seen that the Recombinant DNA technology has allowed the isolation of cDNAs that determine the synthesis of valuable proteins that exist in the human body in minute quantities only. Human insulin and human growth hormone are already being produced in E. coli and sold commercially by Genentech, Amgen has produced a recombinant granulocyte colony stimulating factor (rG-CSF) to fight infections associated with some anticancer drug therapies. The development of DNA micro-injection techniques has spawned the emergence of transgenic animals carrying foreign genes and propagating them to their offspring. In this manner, gene expression can be studied in vivo.

For example, human growth hormone has been successfully transferred to mice, rabbits, sheep and pigs, with the hope of, someday, providing a genetic therapy for human dwarfism. Transgenic cows carrying bovine somatotrophin have increased milk production by 10-25% although the social benefits of this effect are still debated. DNAX Inc., Princeton, N.J., has produced genetically-engineered pigs that synthesise human hemoglobin. Other transgenic animals carrying human genes may provide an abundant source of recombinant therapeutic proteins in the future.

Engineering Vaccines

Diseases that affect a large proportion of the world's population have received high priority for the production of recombinant vaccines. For example, in China, more than 85% of the population has been\exposed to hepatitis B. The gene for the surface antigen HbsAg of the hepatitis B virus was cloned and when expressed in yeast and used as an immunogen, generated a protective immune response. Vaccines are the most important tools in preventing the occurrence of infectious diseases.

Vaccinology occupies an important place in human and veterinary medicine. Edward Jenner, a British physician, invented the idea of injecting small amounts of the fluid containing Cowpox virus into people to immunise them from small pox. The finding of Jenner in 1798 has revolutionised the science of vaccines and vaccines have had profound and long lasting effects on world health. Human medicine has seen the eradication of small pox, polio is nearing elimination and diseases such as diphtheria, measles, mumps, pertussis and tetanus are contained.

Rinderpest in animals is nearing elimination. The need for both new and improved vaccines is even more pronounced in developing countries where effective vaccines are often unavailable or cost prohibitive. This has led to the discovery of new vaccines for well defined human diseases such as measles, polio, tuberculosis, leprosy and malaria, may someday be controlled or even eradicated. There are 5 types of vaccines produced.

— Tissue Culture vaccine

— Purified Protein Subunit vaccine

— Edible vaccine

— Combination vaccine

— DNA vaccine

The cell lines are used for production of various vaccines for animal and human use. Edible vaccines appear to offer the solution to these constraints. There are several reports indicating that antigens derived from various pathogens can be synthesised at high levels in plants. There has been increase in demand for new combinations. Introduction of DNA into the muscle cells of mice resulted in the expression of the genes over extended periods of time and can therefore provide long-lived immunity. This novel vaccine technology is being developed for a variety of clinical applications. Intramuscular inoculation of DNA vaccine offers good protection against rabies virus in mice.

The effect of addition of small quantity of conventional tissue culture rabies antigen to a DNA rabies vaccine was able to enhance the rabies virus neutralising antibodies.

Plant Biotechnology

Genetic engineering of plants is the next green revolution. The technology has already allowed expression of many alien genes into all kinds of plants. This new biology provides powerful tools to augment conventional plant breeding. Plant varieties with good agronomic characteristics can be further improved for better processing, solid content, flavour, colour, texture, acidity, sweetness, size and shape of edible parts. Other characteristics that can be modified include suitability for mechanical harvesting, self storage and resistance to disease, pests, insects and frost.

Applications of Agricultural Biotechnology

Crops: Improvement in yields, ratio of primary to secondary products, nutritive value of products, physical properties—i.e. bread making, resistance to insecticides and pesticides, gene capacity to resist stress.

Livestock: Improvement in disease control, efficiency of reproduction, yields of livestock products, i.e., meat, milk, wool, eggs, composition of livestock products.

Man: Introduction of novel protein sources for man—i.e. meat analogues

Quality and Productivity Improvement in Horticulture

High plant biotechnology has been witnessing major advances in recent years. Some technologists have found extensive use while others are on the anvil or in the developmental stages. The alliance of basic science with the applied sciences in the latter half of the 70s led to the genesis of biotechnology, which has since emerged as a versatile tool packing in powerful techniques for manipulation of life-forms. As the yield falls far short of the requirements of both domestic and export to other developing countries. It is here that biotechnology steps in to envisage enhanced horticultural production.

Another important area in which plant biotechnology is likely to have a considerable impact on the agricultural industry is in the development of plant varieties whose seed proteins have been modified to the needs of man. One such example is the improvement in bread making quality of high yielding, home grown wheat which is largely, though not entirely dependent upon the characteristics of the glutein proteins present in the seed endosperm. A second such objective is to improve the nutritive value of the cereal proteins for man and also pigs and poultry, by alteration in overall amino acid composition of the proteins present; cereal proteins are limiting in lysine and to a lesser extent threonine. It is often overlooked that cereal proteins provide the major dietary source of protein for man.

Salt tolerance and biosaline concept

Salinity is a problem in a very large area of the world. In order to extend intensive cultivation to the marginal environments, the problem is to be faced and it has been reiterated time and again that no solution shall come from an integrated or interdisciplinary approach. In that pursuit information has been generated through working with prokaryotic unicells, eukaryotic algae and higher plants. The genes for production of osmoprotective molecules (osm genes) are present in plants but these have been studied in detail in bacteria. Some of the possibilities are as follows:

1. It is possible to convert osmo-sensitive bacteria into osmotolerant one.
2. osm genes conferring stress tolerance are carried as a segment of DNA comprising about 10,000 base pairs.
3. The segment codes for enzymes catalysing first two steps in proline pathway.
4. Enzyme synthesised through osm gene losses its sensitivity to feed back inhibition by proline, which is the requisite property important for overproduction of metabolites and imparting salt tolerance.

The biosaline concept has proved its relevance in the modern context of biotechnology. The concept was promulgated as poor soils, high solar insulation and saline water, which prevail in lands should be viewed as useful resources rather than as disadvantages, and that these can be used for such non-traditional production of food, fuels and chemicals. Since then much information has been generated on Bio-saline concept. In addition to the application of genetics and molecular biology in developing salt tolerant glycophytes and halophytes. Filamentous blue green alga Spirulina can be cultured on saline water as a major source of single cell protein for animal nutrition.

Protein Engineering

Protein engineering is an important area of biotechnology, through which it is possible to produce superior enzymes and storage proteins, in this area, a protein engineer prepares first a computer aided protein model for a specific function and then prepares a synthetic gene which produces the desired protein in a predictable manner, use of immobilised enzyme has made possible the production of a variety of substances e.g., high-fructose corn syrup (as sweetening agent for soft drink industry), glucose isomerase. Enzyme technology is an area of considerable current interest and development, Enzymes are biological catalysts and have been used for many years as Isolated agents particularly In food industry, e.g., rennin, papain, and invertase.

Microorganisms are very convenient enzyme sources, because the enzymes concentration may be markedly increased by acting on environmental conditions and by genetic manipulation. These enzymes have increasingly replaced plant and animal enzymes: thus, amylases from Bacillus and Aspergillus have substituted those of malted wheat and barley in brewing, and biscuit-making, and in the textile industry, etc. Most of the enzymes are lost in product or the effluent after process, therefore, reusability is essential to reduce the cost of enzyme production and to improve the economics of enzyme catalysed reactions in industry.

Genetically-engineered pesticides

Mycogen's genetically-engineered pesticides (MVP and M-Trak) have been approved by the Environment Protection Agency (EPA) for sale in the United States. To produce these biopesticides, Mycogen inserted the Bacillus thuringiensis (Bt) insecticidal protein gene into Pseudomonas. The recombinant bacteria were fermented and then killed for application to crops. These killed cells encapsulating insecticidal proteins are deadly meal for caterpillar insects that attack cabbage, broccoli, lettuce, and other vegetable crops and against the Colorado potato beetle which damages potato, tomato and eggplant. Monsanto scientists have expressed the Bt toxin gene in the organisms that colonise plant roots. Many plants have been constructed that are tolerant to various insects and pests. Genetic engineering of insects to control crop pests offers attractive possibilities.

Technological Revolution in New Biology

The last decade of the 20th century promises many products for animal and human welfare in the pipeline. Fusion of various metabolic pathways could yield molecules of commercial importance. Now the challenge is the production of complex chemical structures such as taxol and steroid hormones by microbial fermentations. The genetic engineering of the metabolic pathways that involve multiple genes is the next challenge to the genetic engineers. Any investigator can isolate these genes by polymerase chain reaction or obtain them by chemical synthesis with codon optimisation for expression in the targeted organisms.

Sequencing of the Arabidopsis genome will yield the structure of major plant genes. The availability of the sequence of the complete genome of an algae allows isolation of many plant like genes by polymerase chain reaction. However, unique genes involved in secondary metabolism have taxonomically restricted distribution. Biodiversity of organisms is a reservoir of these genes. Isolation and patenting of such genes could be of economic significance. Many unique fatty acids that are source of fragrance, plastic coatings and cosmetics, exist in a number of taxonomically diverse organisms. Some of these genes can be isolated by subtractive hybridisation.

Transgenic crops are a new development in plant breeding. Breeders are no longer restricted to genes in plants that can be crossed with the variety targeted for improvement. The whole living kingdom is now a source of genes and new technology has dramatically expanded the gene pool. The gene transfer is controlled so that only well characterised sequences are introduced. Here are some of the biotechnologically produced medicines, vaccines approved drugs and transgenic crops.

Achievement of Biotechnology

In genetic engineering programmes, it has become possible to map the whole genome of an organism to find out the function of the genes, cut and transfer into another. Owing to the success achieved from gene cloning, many products have been obtained through genetically engineered cells, and hopefully many can be produced during the present decade. Biotechnology has caused a revolution in agricultural science. Cell culture and protoplast techniques have resulted in hybrid plants through inter-generic crosses which generally are not possible through the conventional hybridisation techniques.

For the protection of environment and abatement of pollution, treatment of sewage, transformation of domestic wastes and xenobiotic chemicals have drawn much attention in recent years. To combat these problems such bacterial plasmids have been developed that could be used to degrade the complex polymers into non-toxic forms. Strains of cyanobacteria, green algae and fungi have been developed which could be used for the treatment of municipal and domestic sewage and industrial discharges into nontoxic forms and renew them as source of energy. Biotechnology has helped the bio-industries in producing the novel compounds and optimisation, and scale up products, for example alcohols, acids, antibiotics and enzymes and single cell protein and mycoprotein.

Technologies have also been developed to seek an alternative source of energy from biomaterials generated from agricultural, industrial, forestry and municipal sources. In recent years, techniques have been developed to culture the plant cell, tissue, and organs. In vitro grown cultures (e.g. plantlet, apical meristem culture, etc.) are stored. Storage of in vitro grown cultures has many advantages over the others such as

1. requirement of less space.
2. cheap in maintenance
3. high propagation potential
4. least problem of genetic erosion of stock,
5. maintenance of pathogen free stock.

Immune System

Nonspecific defences are very effective against a wide range of pathogens, but in some cases the pathogens can evade the nonspecific defenses and the body needs an additional defence system for these cases. This is where the specific immune response comes into play. The set of reactions carried out by the immune system following the detection of a foreign invader in a healthy human. These recognitions can occur in the bloodstream or in the lymphatic system, which is a group of vessels similar to veins and arteries that are responsible for draining and cleaning excess tissue fluid (mostly water) before returning the fluid back to the blood.

Monoclonal Antibody Technology

Scientists have always been looking for new ways to use the things that are naturally produced by organisms. Scientists are looking for ways to mass-produce antibodies and use those antibodies to our advantage. In fact, things such as disease diagnosis tests and home pregnancy tests take full advantage of our knowledge of antibodies and how specific they are. A whole area of Biotechnology, called monoclonal antibody technology, is furthering the development of antibody use.

ETHICAL PERSPECTIVE OF BIOTECHNOLOGY

From an ethical perspective, Biotechnology is challenging for three reasons:

— Risks to world views and "slippery slope" risks (series of events that build and move like an avalanche).

— Risks associated with social and economic impacts

— Risks to animal health and human health and the environment ion-style medicine may be a reality.

Following are some possible problems that may result from recent and future developments in Biotechnology:

1. Lowering genetic diversity and overpopulation because of cloning, in other words, reducing the likelihood of natural human evolution which may cause us to be more susceptible to harmful things.
2. Use of genetic information for improper purposes
3. Manipulation of genetic information to create "designer" babies.
4. Genetic testing and genetic determinism (should insurance companies grant/deny coverage based on genetic information?)
5. Improper inferences from genetic information (Genetic discrimination?)
6. The question "Who owns the genetic information?" will be asked and debated.

The HGP may be destined to result in the mass ownership of genetic information and living organisms. Is owning genetically engineered organisms any different from owning non-modified forms of these organisms? Should humans go patent or trademark their genetic material to protect themselves from possible exploitation by biotechnology companies?

2

Biotechniques

Genetic engineering is the most fundamental mechanics of biotechnologies and is a recent off shoot of biotechnological research. It involves gene splicing, recombinant DNA cloning and tissue culture technology. The technique overall involves into two steps: Step one involves synthesis of recombinant DNA and step two is the Gene cloning.

These two, recombinant DNA and gene cloning technologies are the most powerful tools developed in field of biology. Genetic Engineering involves manipulation of the genetic material of an organism to give an altered expression of our choice. It deals with identification and isolation of, desired gene and then joining this gene of interest into another organism. The desired gene expresses itself in that organism by the gene product. Steps involved are:

i) Gene isolation

ii) Selection of vector

iii) Cloning of desired gene

iv) Specific gene transfer

v) Expression of desired gene

Gene Isolation and Cloning of DNA

The formation of new combinations of genetic material by the insertion of nucleic acid produced outside the cell into a virus, bacterial plasmid or any other vector system to allow its incorporation into a host organism in which it is capable of continued replication and expression is termed as genetic engineering. The techniques involved are also referred to as gene cloning, in vitro genetic manipulation or recombinant technology. The transfer of genetic material between individuals takes place in nature by conjugation, transduction and transformation.

Whatever the long-term aim of a molecular biology project the initial objective is usually to isolate a single gene or other specified DNA segment from the organism being studied. This is the essential preliminary step to DNA sequencing, to examine the expression profile of a gene, or for purification of the protein coded by a gene. This initial objective can be achieved in either of two ways:

— By DNA cloning, which involves insertion of the desired DNA molecule into a cloning vector followed by replication of the cloning vector in a culture of E. coli bacteria.

— By PCR, which involves enzymatic amplification of the desired DNA fragment.

'Gene Cloning' is a method of isolating a specific region of DNA and producing millions of identical copies of the DNA (gene) within a microbial cell culture. The basic steps in a gene cloning experiment are:

— Isolation and purification of DNA

— Cutting of DNA in required sites by the same restriction enzymes.

— Joining (ligation) the required piece of DNA to a vector

— Implanting the vector into the host cell by transformation

— Selection of the transformants

— The recombinant DNA molecules replicate within the host cells.

— When the host cell divides, copies of the recombinant DNA molecules are passed on to the progeny.

— Continued replication of the host cells results in clones, each of which consists of identical cells all containing copies of a single recombinant DNA molecule.

This series of manipulations results in a clone library, comprising many different clones, each carrying a different segment of the original DNA. The next step is therefore to identify the clone that contains the gene of interest PCR in outline—PCR is a very different approach to isolate DNA segment. Rather than a lengthy series of manipulations involving living cells, PCR is a test-tube reaction that is carried out simply by mixing together the appropriate reagents and incubating them in a thermal cycler, a piece of equipment that enables the incubation temperature to be varied over fime in a pre-programmed manner.

The basic steps in a PCR experiment are as follows DNA is prepared from the organism being studied and denatured by heating to 94°C. A pair of oligonucleotides is added to the DNA; the sequences of these oligonucleotides enable them to anneal either side of the gene or other DNA segment that is to be isolated, and the mixture is cooled to 50-60°C so that these oligonucleotides attach to their target sites. A thermostable DNA polymerase is added together with a supply of deoxyribonucleotides and the mixture is heated to the optimal temperature for DNA synthesis. The cycle of denaturation-annealing-extension is repeated 25-30 times, with the number of newly-synthesised DNA molecules doubling during each cycle. This exponential amplification results in synthesis of a large number of copies of the DNA sequence flanked by the pair of oligonucleotides. Basic techniques needed for cloning and PCR are:

1. Handling bacteria
2. Preparation of DNA
3. Preparation of RNA
4. Separating DNA and RNA by gel electrophoresis
5. Purifying DNA molecules from electrophoresis gels
6. Construction of recombinant DNA molecules
7. Introduction of recombinant molecules into host cells and recombinant selection
8. Cloning vectors

Handling bacteria: One of the most important practical skills you as a microbiologist is the ability to grow pure cultures of bacteria including cultures infected with bacteriophages. This is because most gene cloning experiments use the bacterium E. coli as the host organism. Even if the longterm objective is to clone a gene in an organism other than E. coli the initial manipulations will be carried out in E. coli because of the ease with which this bacterium can be handled.

Preparation of DNA: For many, the first real molecular biology to be attempted is purification of DNA from the organism being studied.

Preparation of RNA: The preparation of pure, intact RNA is relatively difficult because of the ubiquity of contaminating enzymes that degrade RNA molecules.

RNases present in the cells from which the RNA is being extracted can be inactivated during the purification procedure by a suitable RNase inhibitor, but considerable problems can be presented by glassware and solutions contaminated with RNases derived from skin secretions. These enzymes are very resistant to physical treatments some even withstanding autoclaving and their control can be a problem. RNA can also be used instead of DNA as the starting material for PCR.

Separating DNA and RNA by gel electrophoresis: Separating DNA and RNA molecules of different sizes is achieved by electrophoresis in an agarose or, more rarely, polyacrylamide gel. Agarose gels are routinely run to estimate the sizes of DNA and RNA molecules, which enables the success of a preparative technique or enzymatic manipulation to be assessed. Agarose gel electrophoresis is also the preliminary to analysis of DNA molecules by techniques such as Southern hybridisation.

Purifying DNA molecules from electrophoresis gels: In some areas of molecular biology there are a number of alternative procedures for achieving the same objective.

1. The purification of DNA molecules from gels is one such case. Frequently, an agarose or polyacrylamide gel will provide a band that can be confidently identified as containing a DNA fragment of interest.
2. The next step might be to excise the band and purify the DNA molecules prior to construction of recombinant DNA molecules and further study.

Construction of recombinant DNA molecules: The first real step in a gene cloning experiment is construction of recombinant molecules by insertion of DNA fragments into vector molecules.

Introduction of recombinant molecules into host cells and recombinant selection: Once recombinant DNA molecules have been constructed they must be introduced into E. coli cells. This is quite straight forward and standard technique for different types of vector.

Cloning vectors: By now it should be clear that the choice of cloning vector is an important one. It should determine the nature of the procedure used to introduce the recombinant molecules into the host cells and it should specify the way in which recombinants are selected. It is important to choose a vector that is suitable for the size of DNA fragments that you wish to clone, as some vectors are designed for relatively small fragments (5 kb and less) whereas others require fragments above a certain size.

DNA Sequencing

A DNA sequence on its own is not of much use. If you do not know what to expect then it may not even be possible to assign unambiguously the coding region of a gene. The following are the common methods of DNA sequencing :

— Sequencing

— Manual sequencing

— Automated sequencing

— Southern blotting

— Northern blotting

— Western blotting

Sequencing

One of the major methods of DNA sequencing in known as chain termination sequencing, dideoxy sequencing, or Sanger sequencing after its inventor biochemist Frederick Sanger. The method is elegantly simple. While DNA chains are normally made up of deoxynucleotides (dNTPs), the Sanger method uses dideoxynucleotides.

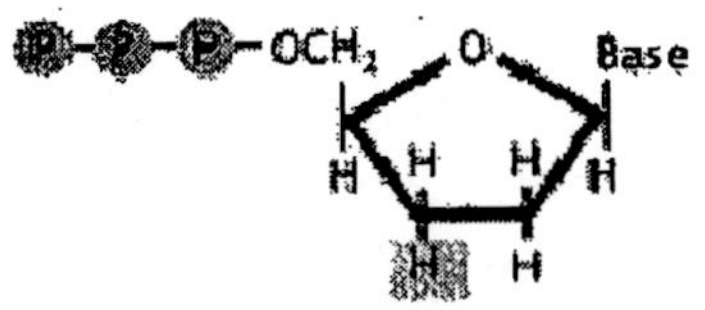

dideoxynucleotide (ddNTP)

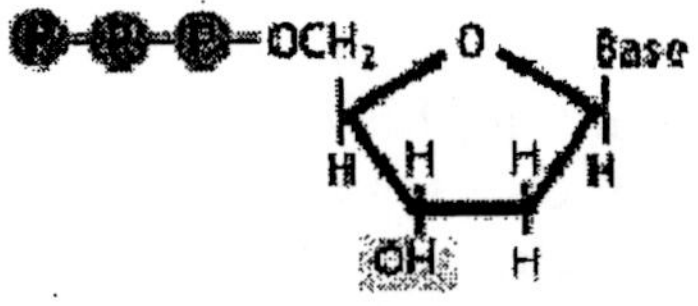

deoxynucleotide (dNTP)

Dideoxynucleotides (ddNTPs) are missing a hydroxy (OH) group at the 3' position. This position is normally where one nucleotide attaches to another to form a chain. If there is no OH group in the 3' position, the additional nucleotides cannot be added to the chain, thus interrupting chain elongation.

Manual sequencing

1. Begin the process by synthesising a chain that is complimentary to the template you want to analyse.
2. Add a specific primer that you know will anneal.
3. Divide your sample among four test tubes. One test tube will be used for each specific nucleotide (dGTP, dATP, dCTP, and dTTP).
4. Add DNA polymerase to each tube and one specific nucleotide per tube.
5. Add ddNTPs to all four tubes. Once you add the ddNTPs, there is no way for the chain to keep elongating, hence you have dd chain termination.
6. Run this on a polyacrylamide gel using one lane per reaction tube (dGTP, dATP, dCTP, and dTTP).
7. To sequence, read the order of bases from the smallest to the largest.

Automated sequencing

An automated sequencer runs on the same principle as

the Sanger method (dideoxynucleotide chain termination). A laser constantly scans the bottom of the gel, detecting bands as they move down the gel. Where the manual method uses radioactive labelling, automated sequencing uses fluorescent tags on the ddNTPs (a different dye for each nucleotide). This makes it possible for all four reactions (dGTP, dATP, dCTP, and dTTP) to be run in one lane, so you can have huge numbers of reactions on one gel. This is a very efficient method.

It is important to remember, however, that a computer can make mistakes. Don't trust the computer. Always check your printout for accuracy. You are looking for a good signal, at least in the 100s, and proper spacing, ideally about 12. Look also for big gaps between the bases since the computer can miss bases. It may often miss a G after an A, especially after an AA.

Southern blotting

Southern blotting was named after Edward M. Southern who developed this procedure at Edinburgh University in the 1970s. To oversimplify, DNA molecules are transferred from an agarose gel onto a membrane. Southern blotting is designed to locate a particular sequence of DNA within a complex mixture. For example, Southern Blotting could be used to locate a particular gene within an entire genome. The amount of DNA needed for this technique is dependent on the size and specific activity of the probe. Short probes tend to be more specific. Under optimal conditions, you can expect to detect 0.1 pg of the DNA for which you are probing.

Northern blotting

Sometimes it's a bit hard to understand, but there is humour in science. In the 1970s, E.M. Southern developed a method for locating a particular sequence of DNA within a complex mixture. This technique came to be known as

Southern blotting. In a tongue-in-cheek fashion, those who used a similar method for locating a sequence of RNA named it Northern blotting. It is also known as Northern hybridisation or RNA hybridisation. The procedure for and theory behind Northern blotting is almost identical to that of Southern blotting, except you are working with RNA instead of DNA.

Western blotting

Western blot analysis can detect one protein in a mixture of any number of proteins while giving you information about the size of the protein. It does not matter whether the protein has been synthesised in vivo or in vitro. This method is, however, dependent on the use of a high-quality antibody directed against a desired protein. So you must be able to produce at least a small portion of the protein from a cloned DNA fragment.

You will use this antibody as a probe to detect the protein of interest. Western blotting tells you how much protein has accumulated in cells. If you are interested in the rate of synthesis of a protein, Radio-Immune Precipitation (RIP) may be the best assay for you. Also, if a protein is degraded quickly, Western blotting won't detect it well; you'll need to use (RIP).

Treatment of Genetic Disease

Until recently, no real thought was given to a treating a genetic disease except for traditional methods of reducing the symptoms of the diseases. With recent developments in gene therapy, the possibility of "correcting" mutated or altered genes has come into being. For example, hemophilia is a disease causes by a defect in the gene that codes for clotting factor. Hemophilia is a disease that causes people to clot blood incorrectly and thus bleed for a possibly dangerous period of time.

The way hemophilia used to be treated was by a complete blood transfusion, which had many side effects, including HIV infection. Now physicians are able to give the clotting factors individually without giving whole blood and without HIV. The main objective of gene therapy is to take cells that have either an "incorrect" gene or a mutated gene, and correct the gene to what is believed to be the normal form of the gene. It is also believed that, even in a partial sense, a human can be created that has the corrected gene.

As far as genetic diseases are concerned today, there is no traditional method available that can provide an effective cure for such a disease. Creating transgenic animals and mice, as you have seen, is becoming pretty commonplace today. The question comes when we are trying to introduce new genes, or at least copies of normal genes, into humans. This is theoretically possible, but many technical and ethical boundaries do exist.

The main problem is the fact that when transgenic animals are created, many individual animals receive the foreign DNA and only the few that survive and display the appropriate qualities are kept. In other words, it's a game of numbers, and it becomes very difficult when you're dealing with an animal that only produces on, maybe two (but there ARE exceptions) offspring at a time. So the bottom line here is that we need to have more effective methods of gene delivery into humans, or else the chance of gene therapy becoming a success in humans becomes very low.

Also, there is a fundamental difference between gene therapy for recessive diseases and gene therapy for dominant diseases. Recessive diseases are caused by the lack of production of a specific product, so gene therapy must provide a specific function. Dominant diseases are caused by an altered function, for which gene therapy must wither return the defective gene to the normal state or inactivate it, along with providing the normal gene. Current technology is far from being able to replace a particular gene, or even to alter the expression of a gene, so we will discuss the treatment of recessive disorders. The possibilities of gene therapy approaches to curing disease are exciting, but are loaded with risks and challenges. However, the prospect of widespread gene therapy brings about many risks, social issues, and ethical issues.

Tissue Culture

Types of Tissue Culture

1. Plant Tissue Culture
2. Types of Plant Tissue Culture

Plant tissue culture, which covers all types of aseptic plant culture should be used as a restricted sense and it is possible to distinguish it into various types of cultures.

- *Seed Culture*: Culture of seeds in vitro to generate seedlings/plants
- *Embryo Culture*: Culture of isolated mature or immature embryos.
- *Organ Culture*: Culture of isolated plant organs. Different types can be DO, the distinguished, e.g. meristem, shoot tip, root culture, anther tissue culture
- *Callus Culture*: Culture of a differentiated tissue from explant allowed to dedifferentiate in vitro and a so-called callus tissue is produced.
- *Cell culture*: Culture of isolated cells or very small cell aggregates remaining dispersed in liquid medium

— *Protoplast culture*: Culture of plant protoplasts, i,e., cells devoid of their cell walls.

— *Anther culture*: Culture of anthers.

Seed culture

Seed culture is an important technique when explants are taken from in vitro-derived plants and in propagation of orchids. Sterilising procedures are needed for plant materials that are to be used directly, as explant source can cause damage to tissues and affect regeneration. In that case, culture of seeds to raise sterile seedling is the best method. Orchid cloning in vivo is a very slow process. Thus seeds can be germinated in vitro and vegetatively propagated by meristem culture is then carried out on a large scale. Most orchids are sown in vitro because: orchid seeds are very small and contain very little or no food reserves. Their small size (1.0-2.0 mm long and 0.5-1,0 mm wide) makes it very likely that they can be lost if sown in vivo, and the limited food reserves also make survival in vivo unlikely.

The seed consists of a thickened testa, enclosing an embryo of about 100 cells. The embryo has a round or spherical form. Most orchid seeds are not differentiated: there are no cotyledons, roots and / or endosperm. The cells of an embryo have a simple structure and are poorly differentiated;

i) sowing in vitro makes it possible to germinate immature orchid embryos, thus shortening the breeding cycle; and

ii) germination and development take place much quicker in vitro since there is a conditioned environment and no competition with fungi or bacteria.

Orchid seeds imbibe water via the testa and becomes swollen. After cell division begun, the embryo cracks out of the seed coat. A protocorm-like structure is formed from the clump of cells and on this a shoot meristem can be distinguished. Protocorm has a morphological state that lies between an undifferentiated embryo a shoot. Protocorms obtained by seed germination have many close similarities with those produced from isolated shoot tips; the term protocorm like-bodies has introduced when cloning orchids by meristem culture.

The vegetative propagation of orchids follows culture of seeds, transformation of meristem into protocorm-like bodies, the propagation of protocorms by cutting them into pieces and the development of these protocorms to rooted shoots. A large number of factors influence the germination and growth of orchids. The mineral requirement of orchids is generally not high and a salt poor medium of Knudson and Vacin and Went are good. Some of the orchids requires (Paphiopedilum ciliolare) for germination while others require low irradiance. Sugar is extremely important as an energy source, especially for those that germinate in darkness. Regulators are usually not necessary for seed germination, and their addition often leads to unwanted effects like callus formation, adventitious shoot formation, etc.

Embryo culture

Embryo culture is the sterile isolation and growth of an immature or mature embryo in vitro, with the goal of obtaining a viable plant. The first attempt to grow the embryos of angiosperms was made by Hannig in 1904 who obtained viable plants from in vitro isolated embryos of two crucifers Cochleria and Raphanus. In 1924, Dietrich grew embryos of different plant species and established that mature embryos grew normally but those excised from immature seeds failed to achieve the organisation of a mature embryo.

They grew directly into seedlings, skipping the stages of normal embryogenesis and without the completion of dormancy period. Laibach demonstrated the practical application of this technique by isolating and growing the embryos of interspecific cross. Linum perenne and L, austriacum that aborted in vivo. This led Laibach to suggest that in all crosses where viable seeds are not formed, it may be appropriate to exercise their embryos and grow them in an artificial nutrient medium. There are two types of embryo culture:

i) Mature embryo culture: It is the culture of mature embryos derived from ripe seeds. This type of culture is done when embryos do not survive in vivo or become dormant for long periods of time or is done to eliminate the inhibition of seed germination. Seed dormancy of many species is due to chemical inhibitors or acids, mechanical resistance present in the structures covering the embryo, rather than dormancy of the embryonic tissue.

ii) Immature embryo culture/embryo rescue: It is the culture of immature embryos to rescue the embryos of wide crosses. This is mainly used to avoid embryo abortion with the purpose of producing a viable plant. The underlying principle of embryo rescue technique is the aseptic isolation of embryo and its transfer to a suitable medium for development under optimum culture conditions. Florets are removed at the proper time and either florets or ovaries are sterilised. Ovules can then be removed from the ovaries. The tissue within the ovule, in which the embryo is embedded, is already sterile.

For mature embryo culture either single mature seeds are disinfected or if the seeds are still unripe then the still closed fruit is disinfected. The embryos can then be aseptically removed from the ovules. Utilisation of embryo culture to overcome seed dormancy requires a different procedure. Seeds that have hard coats are sterilised and soaked in water for few hours to few days. Sterile seeds are then split and the embryos excised. The most important aspect of embryo culture work is the selection of medium necessary to sustain continued growth of the embryo. In most cases a standard basal plant growth medium with major salts and trace elements may be utilised.

Mature embryos can be grown in a basal salt medium with a carbon energy source such as sucrose. But young embryos in addition require different vitamins, amino acids, and growth regulators and in some cases natural endosperm extracts. Young embryos should be transferred to a medium with high sucrose concentration (8-12%); which approximate the high osmotic potential of the intracellular environment of the young embryosac, and a combination of hormones which supports the growth of heart-stage embryos (a moderate level of auxin and a low level of cytokinin). Reduced organic nitrogen as aspargine, glutamine or casein hydrolysate is always beneficial for embryo culture.

Malic acid is often added to the embryo culture medium. After one or two weeks when embryo ceases to grow, it must be transferred, to a second medium with a normal sucrose concentration, low level of auxin and a moderate level of cytokinin which allows for renewed embryo growth with direct shoot germination in many cases. In some cases where embryo does not show shoot formation directly, it can be transferred to a medium for callus induction followed by shoot induction. After the embryos have grown into plantlets in vitro, they are generally transferred to sterile soil and grown to maturity. Applications of embryo culture are:

Prevention of embryo abortion in wide crosses: Successful interspecific hybrids have been seen in cotton, barley, tomato, rice, legume, flax and well known intergeneric hybrids include wheat x barley, wheat x rye, barley x rye, maize x Tripsacum, Raphanus sativus x Brassica napus. Distant hybrids have also been obtained via embryo rescue in Carica and Citrus species. Embryo rescue technique has been successfully used for raising hybrid embryos between Actidinia deliciosa x A. eriantha and A. deliciosa x A. arguata.

Resistance traits transferred to cultivated species through embryo rescue technique.

Crossing species	Resistance trait(s)
Lycopersicon esculentum x *L.peruvianum*	Virus, fungi and nematodes
Solanum melongena x *S. khasianum* (Leucinodes arbonalis)	Brinjal shoot and fruit borer
Solanum tuberosum x *S. etuberosum*	Potato leaf roll virus
Triticum aestivum x *Thynopyrum scripeum*	Salt tolerance
Hordeum sativumn x *H. vulgare* blotch	Powdery mildew and spot
Hordeum vulgare x *H. bulbosum*	Powdery mildew
Oryza sativa x *O. minuta*	Blast *(Pyricularia grisea)* and
Bacterial blight	*(Xanthomonas oryzae)*

Production of Haploids: Embryo culture can be utilised in the production haploids or monoploids. Kasha and Kao (1970) have developed a technique to produce barley monoploids. Interspecific crosses are made with Horeum bulbosum as the pollen parent, and the resulting hybrid embryos are cultured but they exhibit H. bulbosum chromosome elimination resulting in monoploids of the female parent H. vulgare.

Overcoming seed dormancy: Embryo culture technique is applied to break dormancy. Seed dormancy can be caused by numerous factors including endogenous inhibitors, specific light requirements, low temperature, storage requirements and embryo immaturity. These factors can be circumvented by embryo excision and culture.

Shortening of breeding cycle: There are many species that exhibit seed dorm that is often localised in the seed coat and/or in the endosperm. By removing these inhibitions, seeds germinate immediately. Seeds sometimes take up and O_2 very slowly or not at all through the seed coat, and so germinate slowly if at all, e.g. Brussels sprouts, rose, apple, oil palm and iris. H (Ilex) are important plants for Christmas decorations. Ilex embryos remain in the immature heart-shaped stage though the fruits have reached maturity.

Prevention of embryo abortion with early ripening stone fruits: Some species produce sterile seeds that will not germinate under appropriate conditions and eventually decay in soil e.g. early ripening varieties of peach, cherry, apple, plum. Seed sterility may be due to incomplete embryo development, which results in the death of the germinating embryo. In crosses of early ripening stone fruits, the transport of water and nutrients to the yet immature is sometimes cut off too soon resulting in abortion of the embryo. Eg: Macapuno coconuts are priced for their characteristic soft endosperm which fills the whole nut.

These nuts always fail to germinate because the endosperm invariably rots before germinating embryo comes out of the shell. Embryo culture has been practised as a general method in horticultural crops include avocado, peach, nectarine and plum. Two cultivars 'Goldcrest peach' and 'Mayfire nectarine' have resulted from embryo culture and commercially grown.

Embryos are excellent materials for in vitro clonal propagation. This is especially true for conifers and members of Gramineae family.

Germination of seeds of obligatory parasites without the host is impossible in vivo, but is achievable with embryo culture.

Callus culture

Callus is basically a more or less non-organised tumor tissue which usually arises n wounds of differentiated tissues and organs. Thus, in principle, it is a non-organised and little differentiated tissue. The cells in callus are of a parenchymatous nature. When critically examined, callus culture is not homogeneous mass of cells, because it is usually made up of two types of tissue: differentiated and non- differentiated. Explant tissue is a differentiated tissue (roots, stem, leaves, flowers, etc.) which is used as a starting material for callus induction. These explant tissues generally show distinct planes of cell division, cell proliferation and organisation into specialised structures such as vascular systems.

If there are only differentiated cells present in an isolated explant, then dedifferentiation must take place before II division can occur. Parenchyma cells present in the explant usually undergo is differentiation. If the explant already contains meristematic tissue when isolated, then this can divide immediately without dedifferentiation taking place. Callus formation takes place under the influence of exogenously supplied growth regulators present in the nutrient medium. The type of growth regulator requirement and its concentration in the medium depends strongly on the genotype and endogenous hormone content of an explant. These requirements can be put into three categories:

1. Auxin alone (especially in monocotyledons).
2. Cytokinin alone
3. Both auxin and cytokinin (carrot)

If the callus is difficult to induce, or if juvenile callus is needed, then immature embryos or seedlings or parts of these are used. It should be taken into account that type of starting material (juvenile or adult) and the original position of the explant in the plant reflects the endogenous hormone level which have an important influence on processes such as cell division and organ and embryo formation. Many other factors are important for callus formation: genotype, composition the nutrient medium, physical growth factors (light, temperature, etc.).

Sucrose or glucose (2-4%) is usually employed as the sugar source. The effect light on callus formation is dependent on the plant species; light may be required in some cases and darkness in other cases. A temperature of 22-28°C is normally advantageous for callus formation. After callus induction, the callus is grown further on a new medium which referred to as subculturing. When subcultured regularly on agar medium, callus cultures will exhibit an S shaped or sigmoid pattern of growth during each passage. There are five phases of callus growth:

1. Lag phase, where cells prepare to divide.
2. Exponential phase, where the rate of cell division is highest.
3. Linear phase, where cell division slows but the rate of cell expansion increases.

4. Deceleration phase, where the rates of cell division and elongation decreases.
5. Stationary phase, where the number and size of cells remain constant.

Callus growth can be monitored by fresh weight measurements, which convenient for observing the growth of cultures over time in a non-destructive manner. Dry weight measurements are more accurate than fresh weight, but method requires sacrifice of the samples. Mitotic index measurement of cell division rates require extensive sampling to reduce sample error and are not easy to perform.

Organ culture

It is an isolated organ grown in vitro. It can be given different names depends upon the organ used as an explant. For example, meristem or shoot tip culture, root culture, nucellus culture, endosperm culture, ovule culture, a culture for production of androgenic haploids while ovule and ovary culture vitro production of gynogenic haploids. The culture of plant o results in three types of in vitro culture.

Organised: The culture of whole plants (embryos, seeds) and organ has been termed as organised culture. In this, characteristic organised structure of a plant o individual organ is maintained. If the organised structure is not broken down, progeny arise which are identical to the original plant material (e.g. meristem culture).

Non-organised: If cells and/or tissues are isolated from an organised part of a plant, dedifferentiate and are then cultured, a non-organised growth in the form callus tissue results. If the callus disperses into clumps of cells (aggregates) a single cells results, it is referred to as suspension culture. Non-organised culture have very low genetic stability.

Non-organised/organised: This type of culture is intermediate between the above two types. Cells in an isolated organ or tissue, first dedifferentiate and then form tissue which then redifferentiate to form organs (roots or shoots) or embryos. Thus organised structures can develop from non-organised cultures either through techniques or spontaneously. In this the progeny are often not completely identical to the original plant material.

Protoplast culture

Now we will look at the different procedures of protoplast culture. They are as follows:

Isolation of protoplasts: Protoplasts (cell without cell wall) are the biologically active and most significant material of cell. Cooking for the first time isolated protoplasts of plant tissue by using cell wall degrading enzymes viz. cellulase, hemicellulase, pectinase, and protease extracted from a saprophytic fungus Trichoderma viride. Now the protoplasts are cultured in vitro. Sterilisation of the leaf samples with sodium hypochlorite solution.

Rinsing in suitable Osmaticum ie in distilled water or MS medium adjusted to a suitable pH and buffer to maintain osmotic pressure. Plasmolysis of cells takes place by keeping the stripped leaves in 13% mannitol for 3 hours. Peeled leaves are transferred into an already sterilised enzyme solution for 12-15 hours for the facilitation of the enzyme to enter the tissue. Isolation and purification of protoplasts takes place by filtering the enzyme solution containing protoplasts through a nylon mesh (45µm). The filtrate is centrifuged 2-3 times repeating the above steps and finally a specific concentration of protoplast suspension is prepared.

Protoplast culture and regeneration: From the protoplast solution of known density (about 105 protoplasts /ml) about 1ml suspension is poured on sterile and cooled down nutrient medium in Petri dishes. The plates are incubated at 25°C in a dim white light. The protoplast regenerates a cell wall, undergo cell division and forms callus. The callus can be subcultured. Embryogenesis begins from callus when it is placed on nutrient medium lacking mannitol and auxin. The embryo develops into the seedling s and finally into mature plants.

Anther culture

Anther culture is the process of using anthers to culture haploid plantlets. The technique was discovered in 1964 by Guha and Maheshwari. This technique can be used in over 200 species, including tomato, rice, tobacco, barley, and geranium.

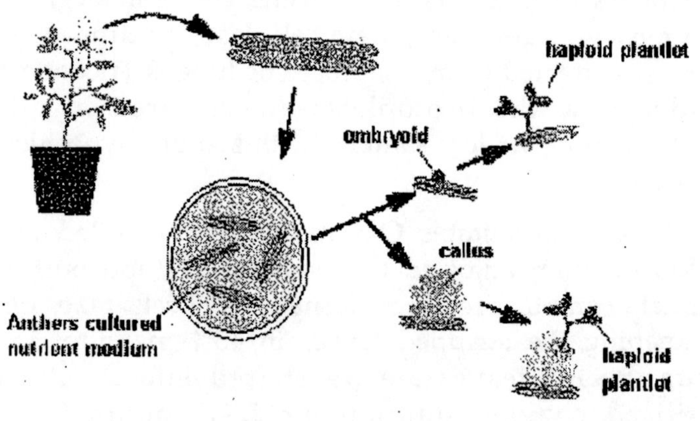

Anther Culture

Some of the advantages which make this a valuable method for obtaining haploid plants are: the technique is fairly simple it is easy to induce cell division in the immature pollen cells in some species a large proportion of the anthers used in culture respond (induction frequency is high) haploids can be produced in large numbers very quickly. In experiments using Datura innoxia, induction frequencies of almost 100% and a yield of more than one thousand plantlets or calluses have occurred under optimal conditions from one anther. Success can be determined within 24 hours as cells begin to divide.

Methods Tissue Culture

The setting up of a tissue culture laboratory needs proper planning. It depends upon space availability, volume of work to be carried out and funds. Usually, the available laboratory space, is be divided into five distinct laboratory areas. These are: Media preparation area/room, Aseptic transfer chamber area, Environmentally controlled culture room, Analytical room and Acclimatisation room.

Media preparation area

Laboratory tables or benches and revolving stools with adjustable height, Hot plate and magnetic stirrers, analytical loading single pan balance-with precision of 0.001 g, weighing range 0.1 mg- 80g, digital read out, Top pan loading balance for quick weighing, range 100 mg- 500g capacity, sensitivity 0.1 g, Refrigerator and freezer, Water purification and storage system, Glassware washing facility with proper drainage, Gas outlet, Electric hot-air oven range upto 250 °C and microwave oven, Digital pH meter, Range 0-14 pH, accuracy 0.1 pH, temperature compensation 0- 100°C, Autoclave preferably horizontal., Continuous supply of single and double distilled water.

Storing glasswares, plastic wares, chemicals, plugs and appliances required for media preparation are as follows:

Requirements: Culture tubes/conical flasks/petri dishes of various capacities, Measuring cylinders- 25 ml, 100 ml, 500 ml, 1000 ml, Cotton for plugs/plastic caps (autoclavable), General glasses/plastic wares of various capacities such as volumetric flasks, beakers, reagent bottles, pipittes, vacuum filtration system and glass rods.

Washing Powder/ Liquid Detergent, Disinfectants: Powder or liquid detergents or wetting agents such as Tween-20, 70% alcohol and absolute alcohol). Glass distilled water, Stock solutions of nutrients of tissue culture media or readymade/ prepared powdered media, Sucrose, Agar (tissue culture grade), Sterile culture vessels with distilled water. Chemicals of analytical grades (Inorganic, organic salts, vitamins, amino acids, growth tors/hormones and activated charcoal), Coconut milk, yeast extract, malt extract, casein lysate and extracts of potato, carrot and tomato. Spatulas, weighing butter paper/boats, stirring bars (magnetic) for magnetic stirrers and stirring rods, Brush (flask and test tubes), gloves (disposable 23 cm), mop (household), scoop, towels Centrifuge (low speed): Variable electronic speed control speed indicator, Amp. meter, timer, dynamic break, starting switch, 230V 50Hz. Electric Autoclaves-Vertical: with safety valves, pressure gauge, steam-release cock.

Horizontal: mounted to tubular stand heavy hings

Steamers: With immersion heaters of ejection safety device, size: 31 cm x 13cm x 10cm to 61cm x 20cm x 15cm.

Filter Sterilisation Equipment: Syringe filter holder: 2.5cm -Pressure Filter Holder: 4.7cm Double Distilled Water Equipment: Built-in energy regulator all glass water stills, heat re-sistant boiling flask with heater.

Environmental Growth Cabinets: Cabinet with controlled temperature, light and humidity, temperature range 4° to

45°C with timer to regulate photoperiods. Gyratory Shakers: Capacity: 50 to 1000 ml flasks, speed: 80 to 200 rpm.

Laminar Air-flow Cabinets: Constant flow of purified air, sizes: 0.6m, 1.2 m and 1.8m. The other miscellaneous equipments which are required for tissue culture are: air conditioners, arrow heads, bunsen bummers, deep-freeze, dissecting needles, glasswares, forceps, fluorescent lamps/ tubes, heaters, hot plates with magnetic stirrer, inoculation cabinets, metal trays and bowls (for transport of cultures), tubes, refrigerators, UV germicidal lamps, wooden or metal racks, etc.

Aseptic technique

In in vitro condition plant cells and microbes have basically same requirements. When the culture medium contains sugar (as carbon source) it attracts a variety of microorganisms which grow fast than that of the cultured tissue in medium and they ultimately kill the plant cells. It is, therefore, necessary to have complete aseptic condition around the culture equipments which prevents contamination of the culture medium. Following are the main, sources of contamination of the medium and the subsequent methods to check them:

1. The microorganisms may be present in the nutrient medium at the time of its preparation. These microorganisms can be destroyed by proper plugging and autoclaving the culture tubes/flask. The medium can be completely sterilised by maintaining it at 120° C for about 20 minutes at 15 Ib pressure in the autoclave.
2. The explant (plant part to be cultured) may carry microorganisms with it, therefore, the excised part should be surface-sterilised by mercuric chloride (1 to

2%) or by sodium hypochlorite solution for 30 minutes.

Precautions must be observed to prevent the entry of microorganisms when the plug of the culture is removed during transfer of the plant material to the medium or from one medium to another. The inoculation chamber may be sterilised by UV -radiations. Correct pH of the medium is important. Highly alkaline or acidic pH affects the nutrient up-take in culture tissues. Therefore, the tissue culture medium is adjusted to a pH of 5.6 to 6.0, before autoclaving.

Semi-solid and liquid media are most commonly used for growing plant cells. A high concentration of gelling agent (agar-agar, gelatin, silica gel) makes the medium very hard and decreases the nutrient uptake by the tissues. Agar at 0.8% to 1,0% concentration is widely used. It is essential to remove dirt and debris from the plant tissue and should be washed in a weak detergent solution and rinsed several times with distilled water prior to sterilisation, Some woody trees, such as buds arid twigs, are cleaned by immersing them briefly in a 70% ethanol solution, mulch wets and spreads over the tissue surfaces more effectively than a higher concentration alcohol.

Sodium Hypochlorite (NaOCI): It is the most common chemical agent used to sterilise plant tissues (0.025%-0.25% NaOCI). Diluted house hold bleach can also be used for this purpose, which normally contains 5,25% NaOCl. It is equally effective and considerably less expensive, Calcium hypochlorite (CaOCI): It can be used as a substitute for NaOCI. CaOCI causes slightly less damage to plant tissues but tends to precipitate out of solution. To avoid the accumulation of CaOCI on the plant tissue surfaces, sterilisation solutions should be filtered or decanted prior to use.

Hydrogen Peroxide (H20V Solution: Plant tissues can also be surface sterilised using a H_2O_2 (3%-10%). It is much easier to remove from tissues than NaOCI and CaOCI:

Other Substances: Plant tissue can also be surface-sterilised by bromine water (1%-2%), silver nitrate ($AgNO_3$ I %) and mercuric chloride ($MgCI_2$ 0.1 %-I %),

Cleaning: Clean the glasswares/plastic wares in 10% commercial detergent liquid or powder for I hr and then in HCI for 2 hr. Remove the traces of detergent and acid by thorough washing with tap water. Rinse vessels with double distilled water and allow them to dry over night at room temperature. Plug glasswares such as conical flasks or test tubes with non-absorbent cotton or cover by the plastic caps, Wrap the petri dishes with aluminium foil. Place forceps and scalpels in test tubes, plug the tubes with cotton or cover with aluminium foil. Plug the mouth end of the pipettes with cotton. Wrap them individually in aluminium foil.

Animal Tissue Culture

Isolation of animal material (tissue): Before culture, animal materials must be washed in BSS (balanced salt solution) aseptically to avoid contamination. The tissue to be cultured should be properly sterilised with 70% ethanol, and remove surgically under aseptic conditions. Thus the tissue isolated should be either stored in the freeze or used immediately. The following are the various stages of culturing the tissue.

1. Disaggregation of the tissue- There are two methods of disaggregating the tissue: Physical (mechanical) and Enzymatic disaggregation.

   a) *Physical disaggregation*: The aseptically isolated tissue is kept in a sieve of 100μm sieve. It is put in a sterile Petri dish containing buffered medium with balanced salt solution. The cells are alternately passed through sieves of decreasing pore size (50µm and 20µm mesh). The disaggregated cells are now counted using a haemocytometer. The viable dissociated cells are now termed as "Primary cells". When the primary cells are grown on culture medium in high density, this culture is known as primary adherent cell culture.

b) *Enzymatic disaggregation*: Enzymes are also used for dislodging the cells of tissue. By using enzymes a high number of cells are obtained. Disaggregation of embryonic tissue occurs more readily than that of adults or new borns. There are two important enzymes used in tissue disaggregation—collagenase and trypsin.

Use of collagenase: Collagenase is used for disaggregation of embryonic, normal as well as malignant tissues. The intracellular matrix contains collage, therefore collagenase disaggregates normal and malignanat tissues.

First the biopsy tissues are kept in medium containing antibiotics. Thereafter the tissue to be disaggregated is dissected into pieces in BSS containing antibiotics. The chopped tissue after washing is transferred incomplete medium containing collagenase. After five days of treatment the mixture is pipetted so that the medium may get dispersed and the epithelial cells are settled. The settled cluster of the epithelial cells are washed thoroughly and the cell suspension is made free from the enzyme collagenase by centrifugation. The suspension consists of enriched fibroblast fraction which is plated out on medium. The cells grown are the primary cells and the culture is the primary cell culture.

Genetic engineering has enabled scientists to transfer genes across species, families, and even kingdoms. These boundaries, once thought of as impossible in the past, are now readily crossed with modern gene transfer methods that allow the transfer of DNA into living cells and fertilised eggs of animals. Transgenic animals not only provide research tools for studying new gene regulation and disease, but they may be genetically modified for the production of pharmaceuticals, vaccines, and rare chemicals as well as for food production.

Animal Gene Transfer

One challenge in creating transgenic animals is making sure that the transgene is turned on at the right time and in the right tissue. The integrated gene must be expressed and regulated appropriately (for example, in a tissue-specific manner). So the gene needs to be accompanied by the correct promoters and regulatory sequences such as enhancers. Also, the gene must be in as close to the correct location in the genome as possible. The construct must then also be incorporated into the chromosome where gene expression is regulated. Two methods of introduction of genes into cells are microinjection and embryonic stem cell gene transfer.

Microinjection

This is generally the method of choice for creating transgenic animals. Once a gene has been characterised and expressed in eukaryotic cells, a transgenic animal can be made by microinjection of the cloned gene into the fertilised egg of a donor animal. The key here is that the foreign gene must be injected into the cell before the first cell division, so that all cells of the organism contain the desired gene. Donor females are superovulated (produce multiple eggs) and mated, and the fertilised eggs are removed and placed in a sterile dish with buffer.

DNA is injected into the eggs and are implanted into the oviduct of a female mouse made to have many of the symptoms of pregnancy by hormones. After birth, to find which of the offspring are transgenic, animals are subjected to tissue analysis by Southern blotting or PCR. Founder animals, which are animals that have the foreign gene within their germ line (sex cells) and the somatic cells, are bred to establish new genetic lines with desired characteristics. However, there are problems because few injected eggs survive, and not all of those retain the new DNA.

The new DNA is integrated at random if it is not targeted to a specific chromosomal location. Microinjected genes often integrate into a single chromosomal location, regardless of whether the location is used a lot or not used very often. Also, many transgenic animals are mixtures (mosaics) because the gene is only in some of the cells. In order for this to be effective, all of the somatic cells, as well as the germ cells, need to have the foreign gene.

Embryonic stem cell gene transfer

Embryonic stem (ES) cells are also used for producing transgenic animals. ES cells are obtained from the inner cell mass of donor blastocysts (fertilised eggs that have undergone several cell divisions to for a hollow ball of cells called the blastocyst). The cells are cultured prior to transfection with a foreign gene. A gene is targeted to specific areas by homologous recombination and the use of selectable markers (antibiotic resistance). Treated cells are then screened either by PCR or another selection method such as antibiotics.

Transformed ES cells are microinjected into blastocysts so that they can become established in the somatic and germ line tissues. They are then passed on to successive generations by breeding founder animals. At present, the ES method is most successful with mice because mouse ES cells are pleuripotent (can differentiate into many cell types) and, when integrated into blastocysts, can divide and differentiate in the mouse embryo. However, today the cells do not need to be pleuripotent when placed in blastocyst embryos.

The mouse is used as a model in the study of gene function, and is useful in many areas of medicine. Methods are used to inactivate gene function and look for changes in the phenotype, and gene targeting can also be used to add genes encoding desirable characteristics. ES cell transformation technology has been used to inactivate specific genes in order to determine their effect on growth and development or to develop effective treatments.

Gene targeting

The insertion of DNA into a specific chromosomal location, and is often used to inactivate a specific gene into the genome. It occurs in the following steps:

1. A specific gene is isolated, altered according to the requirements of the experiment, and then targeted to its counterpart on a specific chromosome in cells. A part of the gene is inactivated by inserting an antibiotic resistance gene into it, with regions of the original gene flanking the marker remaining on either side. This marker allows cells with the gene to be selected. This is called positive selection. A second marker (called negative selection), usually the herpes simplex virus type I thymidine kinase (HSV-tk) gene, is located at the end of the targeting vector. This marker is used to select for cells in which the transferred gene is targeted to the specific site, rather than a random site or nowhere at all. The gene construct is assembled and the vector is placed into an embryonic host cell, where homologous

recombination allows the gene to insert into a specific chromosome location. Ideally, the target gene, antibiotic resistance gene, but NOT the thymidine kinase gene, are transferred to the chromosome. The tk gene is not transferred because it is out of the regions of the regions that we want to enter the chromosome.

2. Selection of cells: You want to select for cells with the foreign gene. Cells are placed in a medium containing antibiotic (usually neomycin), which kills cells where the gene is not inserted into the genome. Also, a second drug called gancyclovir selects for targeted insertion by killing any cells which contain the tk gene (remember that the tk gene is something that we don't want to enter the cell). What we want is the antibiotic resistance gene entering the cells (so the antibiotic doesn't kill the cells), and the tk gene to NOT enter the cells (or else the gancyclovir will kill them). If the cells have undergone homologous recombination, they will survive treatment with neomycin and gancyclovir. If we get the correct ES cells, then they are placed into mouse embryos, where they become incorporated into the genome. These genetically manipulated embryos are placed in surrogate mothers and develop into young.

3

Benefits of Genetic Manipulation

Novel biotechnology research is rapidly expanding with a view to the application of environment friendly bioprocesses for chemical, pharmaceutical, and other areas, including food bioengineering and safety. The tremendous growth of the two areas of modern biotechnology (i.e., agricultural and medical) may have recently created the impression of technological isolation through specialisation of both agricultural and medical biotechnology. Unfortunately, about one billion people in the world remain deficient in several vitamins and minerals intake. Such undernourished populations live mostly in developing countries that often have inhospitable climates which favour low efficiency in food engineering and production, and also where the primary services for storage or administration of GMO-based vaccines are minimal.

The utilisation-in combination with agronomical techniques-of biotechnological tools that were originally produced for the medical area, may help resolve such difficult problems. As an example, the production of vaccines from plants and microorganism with subsequent addition in foods could reduce the cost of vaccination in such areas. The modification of microorganisms for the bio-production of fuels is also important, and would also be important in under-developed, or developing, countries that have a large surplus of under-utilised complex carbohydrate sources that could be used for bioproduction of valuable fuels with existing technologies.

Immobilisation of Cells

Immobilisation of whole cells has been defined as the physical confinement or localisation of intact cells to a certain defined region of space with preservation of some, or most, catalytic activity. The increased stability under extreme conditions of pH and temperature, as well as the re-use and applicability in continuous processing systems that enclose immobilised cells instead of soluble enzymes make the cells a preferred, versatile tool in both food industry and medicine. There are several different approaches to the classification of immobilised biocatalysts, but the most frequently employed classification is based upon the method of immobilisation selected for a specific application.

The selection of immobilisation method depends, therefore, upon the application, the nature of the microorganism being immobilised, as well as the resources available. Table 1 shows several, possible immobilisation methods that available for whole microbial cells. Most immobilisation methods can be applied either to whole cells or to enzymes. Some of the advantages of whole-cell immobilisation in comparison with enzyme immobilisation are: the higher stability and enzyme activity, multivariate enzyme applications, and the lower cost. On the other hand, disadvantages of using whole cell immobilisation in comparison with enzyme immobilisation are linked to the increased diffusional barriers caused by the much larger sizes of cells in comparison with enzymes.

Table 1. Immobilisation Methods for Whole Microbial Cells.

		Flocculation
	Weak bonds	Adsorption
Binding		Ionic
	Strong bonds	Covalent
		Cross-linking
		Thermal Gelation
	Entrapment	Ionotropic Gelation
Physical retention		Polymerization
	Membrane Retention	Dialysis Culture
		Ultrafilters

Adsorption is the least expensive and mildest immobilisation method. It uses weak interaction forces such as hydrogen bonds, hydrophobic interactions and Van der Waal forces to immobilise cells or enzymes. However, the sensitivity of this interaction to pH makes the leakage of cells immobilised by this technique quite common. Important applications of this technique are related to the production of fructose and vinegar, and also waste water treatments.

Ionic binding uses the properties of negatively charged microbial cells to interact with positively charged ion exchangers. The results obtained with this technique are also sensitive to extreme pH values, and the binding strength is greater in comparison with adsorption. However, the mild conditions employed by this technique make it suitable for use for immobilisation of both enzymes and whole cells. Covalent binding and cross-linking offer better strength than the previous techniques, however, there is an encountered toxicity in the reagents that are used to produce immobilisation. Entrapment techniques are, however, the most commonly encountered in the industry and they are based on the formation of thermally reversible gels, ionotropic gels and polymerisation.

Application Immobilisation

Entrapment of cells in a gel-like matrix by ionotropic gelation using alginates and Kcarrageenans is certainly the most useful method for industrial purposes. The properties of the gel-like matrix allow the cell to remain viable and with its catalytic ability for a long period of time. For example, an increment in the yeast concentration obtained through immobilisation techniques has helped the brewing industry to reduce fermentation process times and the size of their storage facilities. Unfortunately, because of the high concentration of diacetyl, and the low concentration of higher alcohols and esters, the flavour of the fast fermented beer has been compromised.

Even the amino acid profile has been altered. The main factor causing this uncommon imbalance is the insufficient mass transfer in the older designs of fermentation reactors; thus, the use of new reactor designs, and combining technologies could improve the quality of the products obtained through such fermentation reactions. Nutraceuticals are defined as food components that have health benefits beyond traditional nutritional value.

Novel biotechnology tools like immobilisation were also applied for the isolation and incorporation of such food components in ordinary foods. The synthesis of nutraceuticals was reported to be successful by empoloying immobilised lipases, such as those from Candida antartica and Lactobacillus ruteri. The introduction of conjugated linoleic acid (CLA) in dairy foods has been made possible through the immobilisation of lipases.

There is a quite extensive list of immobilisation technique applications in medicine. A very important group of such applications is concerned with the regulation of equilibrium between coagulation and dissolution of coagulated blood (fibrinolysis) through the use of immobilised enzymes. The high probability of death caused by thrombosis (involving the formation of clots in the blood vessels), has committed physicians to the use of fibrinolytic therapy for the treatment of occlusions in those parts of the body where a surgical intervention would be too risky. Amongst the most important enzymes that have been immobilised for use in such therapy are Plasmin and Heparin. The use of biotechnology as well as microscopic techniques has helped refine and greatly improve such therapeutical means.

Current regulations for the disposal of toxic chemicals in the environment as well as the detoxification of water used in any agricultural and industrial process brings the need for novel biotechnology tools to be developed in order to solve such problems in a cost-efficient manner. Enzymes have been isolated from genetically manipulated microorganism strains with the purpose of accelerating the rate of degradation of organic and some inorganic compounds in wastewater as well as in soils.

Genetic Manipulation Techniques

Genetic manipulation techniques have played an important role in enhancing the performance of microorganisms that are significant in the industrial, pharmaceutical and medical application areas. It is important to highlight the fact that they are generic, and also that they can be employed to improve the performance of any cell or microorganism for specific applications.

Mutagenesis

Mutagenesis can be defined as the group of techniques or processes that produce changes in a DNA sequence which modifies either the expression of genes or the structure of the gene products. Two subcategories of mutagenesis are usually applied in the genetic manipulation of microorganisms: classical mutagenesis and transposon-directed mutagenesis. Classical Mutagenesis involves the use of chemical mutagens to regulate a process of interest in the target microorganism.

The main purpose of this 'classical' approach is to find the mechanism of action of the targetted microorganism. Transposon-directed Mutagenesis consists in the genetic alteration of a few sites on the target chromosome in order to determine the function of this specific site of the chromosome. By comparison with the previous category transponson-directed mutagenesis it is more specific, and is therefore preferred in specific applications that require higher selectivity.

Gene Transfer

Electroporation involves the production of transient pores in a bacterial cell membrane following the application of high-voltage electric field (DC) pulses of short duration. Such pores allow the introduction of DNA into the cell under certain, favourable conditions. Such methods have been used with various microorganisms and even plasmids that are of significant use in the industry. Another technique that is employed for the transference of genes involves Shuttle Vectors. Shuttle vectors are DNA constructs that are able to replicate and deliver DNA to widely divergent types of bacteria. DNA transference is obtained after the use of a microorganism whose mechanisms of action have been thoroughly studied, as for example, is the case of various E. coli strains.

Gene Cloning

Gene cloning and genomic maps to clone the genes of

industrially-significant microorganisms have also been reported to be subsequently introduced in native strains of the microorganism of interest.

DNA Analysis

DNA analysis involves the study of a particular region of DNA by either physical or chemical techniques. It is usually carried out in order to obtain insights into the genetic organisation, the mechanisms of gene expression, or to obtain information that may be useful for the construction of shuttle vectors.

Applications to Food Industry

Pioneering studies of genetic manipulation of microorganisms with significant applications to foods were initially based on Escherichia coli as a microbial model because of its ease of growing and manageability. Subsequent studies were carried out in the production of dairy starters, as well as the amino acid and yeast production for brewing purposes, and more recently attempts are being made to produce butanol through fermentation.

Epigenomics

Upon completion of the US Human Genome Mapping Project and related studies, it became increasingly evident that a sequence of 30,000 or so 'active' genes that encode and direct the biosynthesis of specific proteins could not possibly exhaust the control mechanisms present in either normal or abnormal cells (such as, for example, cancer cells). This is even more obvious in the case of developing embryos or regenerating organs. Recently, specific control mechanisms of cellular phenotypes and processes were proposed that involve epigenetic controls, such as the specific acetylation<—> de-acetylation reactions of DNA-bound histones.

Epigenomic tools and novel techniques begin to address the complex and varied needs of epigenetic studies and their applications to controlling cell division and growth. Such tools are, therefore, potentially very important in medical areas such as cancer research and therapy, as well as for improving 'domestic' animal phenotypes without involving genomic modifications of the organism. This raises the interesting question if epigenomically controlled-growth organisms (ECGOs)—to be produced in the future-would be still objected against by the same groups of people that currently object to GMOs, even though genetic modifications would be neither present nor traceable in such organisms?

MICROENCAPSULATION

Microencapsulation is now being extensively used in the industry and medicine to enhance the stability and shelf life of products, to improve the nutritional value of foods, and to produce enzyme, or immuno-protective, capsules for the production of vaccines and for the transplant of organs. However, the prospects of microencapsulation techniques are not limited to these areas because it can be used in combination with other novel biotechnology tools to broaden the spectrum of its applications. The concept behind microencapsulation involves surrounding a component of interest (such as vitamins, proteins, flavour, etc.) that need protecting with a (μm-thick) thin layer of either polysaccharide or lipid material in order to prevent, or significantly reduce, the deterioration of the encapsulated component.

As a result, liquid droplets, solid particles or gaseous materials are packaged with continuous shells that can release their contents at controlled rates under desired conditions. There are several, 'classical' techniques that are used for microencapsulation such as: spray drying, molecular inclusion and extrusion. On the other hand, liposome microencapsulation is a novel technique that is being developed to the point that it can introduce any substance of interest in an organism regardless of the substance's solubility, electrical charge, molecular size or structure.

Liposomes and Liposome

A liposome can be defined as an artificial lipid vesicle that has a bilayer phospholipid arrangement with the head groups oriented towards the interior of the bilayer and the acyl group towards the exterior of the membrane facing water. Liposomes are usually made of phosphatidylcholine (lipid) molecules although mixtures of phospholipids can also be employed to make liposomes. The main difference between liposomes and cell membranes is the lack of bilayer asymmetry in the case of liposomes. On the other hand, liposomes do exhibit barrier properties very similar to those of cellular lipid membranes, and therefore, they are frequently used as a model system for the study of the transport activity of the cell membrane boundary.

Since the invention of liposomes in the early 1960's as 'membrane-like' structures, they have been extensively used in the biotechnology field. The 'membrane-like' characteristics, their colloidal size as well as their solubility properties for various molecules have made liposomes useful tools for various biotechnology applications, from foods applications (such as in enzyme microencapsulation), to fermentation and several medical/ pharmaceutical applications.

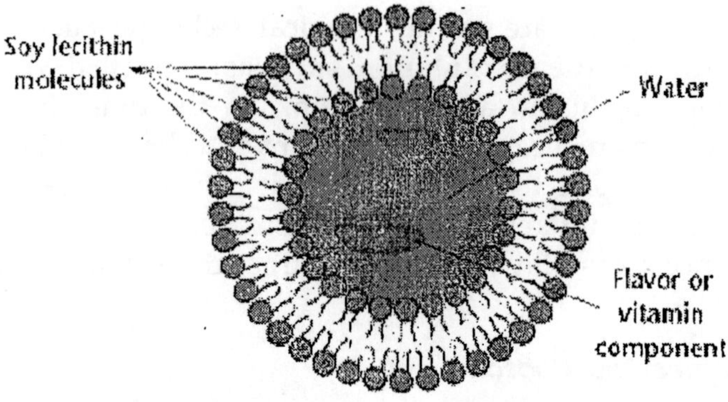

Figure 1. Simplified representation of the molecular organisation of a liposome microcapsule in water.

A few possible mechanisms for liposome absorption into cells are illustrated in Fig 2. Fusion and endocytosis are the most common ones being utilised for the production of liposomes with industrial or medical purposes.

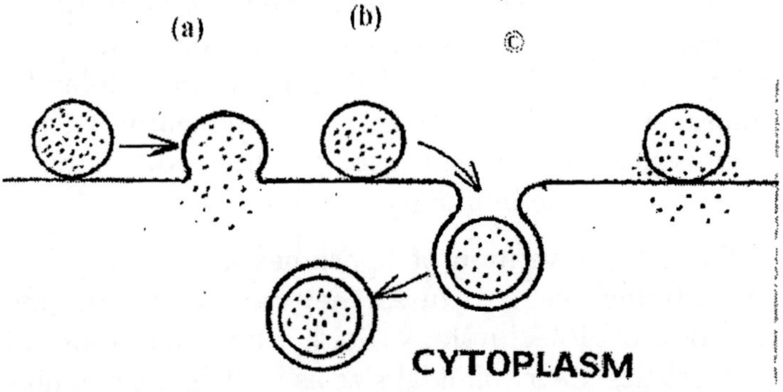

Figure 2. Possible interactions of a liposome with a cell membrane: (a) Fusion; (b) Endocytosis; (c) Adsorption.

During fusion, the soluble material carried in the liposome is released into the cytoplasm of the target cell through the joining of the liposome with the cell membrane lipids.

Such a mechanism is useful for introducing hydrophobic material into the cell, and it has been already employed by the petroleum industry. In the case of endocytosis, the liposome enters the cell surrounded by an endocytotic vacuole that is derived from the membrane lipids. The contents of the liposomes are then released by the action of liposomes that were attached to the produced vacuole. Low molecular weight compounds (that are not charged at low pH values) will thus be able to diffuse into the cytoplasm.

Classification of liposomes

Liposomes are classified according to their size and structure, the latter also depends on the preparation method. There are three classes of liposomes: multilamellar vesicles (MLV's), small unilamellar vesicles (SUV's) and unilamellar vesicles. The type of liposome being utilised in a specific application depends on the hydrophobicity of the molecule being encapsulated. MLV's consist of a series of multiple lipid bilayers that are obtained from a phospolipid solution in chloroform, which is then evaporated to produce a thin film and subsequently hydrated in an aqueous solution to form heterogeneous vesicles with diameter sizes from ~0.3 to 2.0 µm.

Table 2. Characterization of Liposome Properties by Various Techniques.

Electron Microscopy	Liposome Size distribution Average size of Liposomes
Radiactive Tracers	Liposome stability vsus time and temperature Encapsulated Molecule Retention rates
Fluorescence Quenching	Liposome stability Liposome-cell interactions
Ultrasonic Absorption	Temperature induced transitions Limitations of Liposome uses in the presence of hydrophobic proteins

Electron Spin Spectroscopy	Permeability of liposomes Molecular dynamics of Spin-Labelled Lipids in Liposomes and Encapsulates
Nuclear Magnetic Resonance	Molecular dynamics of Lipids in Liposomes and Encapsulates Water exchange in phospholipid vesicles Stability using metal cations Storage of Liposomes
Fluorescence Correlation Spectroscopy	DNA interactions with molecules encapsulated within liposomes Hybridization of DNA for Liposomes Vaccines

The main advantage of the MLV's is that both the lipids and the aqueous solution to be encapsulated are not subjected to harsh treatment. The use of high-intensity ultrasonication, ethanol injection and pressure applied to multillamelar vesicles allows the production of single bilayer vesicles SUV's; this physical treatment reduces the liposome size. Unfortunately, the smaller size of such a vesicle also results in a higher curvature that limits its capture volume.

LUV's are prepared commonly by infusion, reverse-phase evaporation and detergent dilution methods. They vary in size range from 100-500 nm and they are the most widely employed liposomes because they have less variable size and higher entrapment volumes (>~2.7 L / mol of lipid) than SUV's. Several techniques may be employed to characterise liposomes. Table 2 compares some of the current techniques and novel approaches to the study of liposomes and their interactions.

Applications of liposomes

Lecithin-based liposomes offer great flexibility and shelf life improvements in the food industry for introducing water-soluble substances such as flavours and micronutrients. Such modified liposomes are also being used at present in the beverage and bakery industries to deliver both flavours and oils that are trapped inside the liposomes; such flavours and oils are released then into the mouth; these are also employed to incorporate flavour oils. The major impact of such techniques has been achieved through the use of microencapsules that can be made through a continuous process on an industrial scale.

In the dairy industry, liposomes containing enzymes have been reported to reduce the ripening time by 30-50%, as well as improve texture and flavour. The latter was caused by a decrease in the action of proteolytic enzymes in the early phases of cheese fermentation. Because liposomes have the ability to carry fat-based flavours in their bilayer, as well as water-soluble flavours in the core of the vesicle, they protect the flavour from degradation and also increase the longevity of the flavour in the system where they are being employed. Therefore, their use in the beverage industry has become widespread.

The rate of diffusion through/from the bilayer depends on the liposome composition as well as physical properties of the flavour. Bakery is another area where liposomes have been applied and it is based on the characteristic of the liposomes of not being destroyed during the processes of mixing or extrusion; therefore, they can release encapsulated flavorings, fragrances or food additives. When a flavour is encapsulated, the release occurs after the enzymatic degradation of the liposome, and thus the rate of release depends on the physical properties of the material of which the liposome is made. In the case of lecithin, the pH value or range, as well as temperature, are important factors.

Traditionally, liposomes have been used in the bioengineering field to over-produce certain proteins through the genetic modification of cells. They have thus solved the inconvenience of transferring high molecular weight molecules through cell membranes. The use of unilamelar vesicules made of cationic lipids has improved the transfection efficiencies and prevented interactions with DNA molecules. Therefore, the project of introducing genes to cure diseases (genetic therapy) is not far from becoming a reality if the patient being treated were found not to suffer from severe side effects.

Vaccine formulations based on liposomes have been successfully tested in animal immunisation, and such studies are currently in the clinical testing phase. The benefits and limitiations of liposomes as drug carriers in a system as complex as the human body depend basically on the interactions of liposomes with the cells, as well as their immuno-compatibility, or their ability to escape detection by the human immune system.

Cosmetics is another area where the liposomes have been extensively employed. They are being utilised as humectants, as well as carriers of formulations containing extracts, vitamins, moisturisers, antibiotics and proteins. Such applications are mostly directed towards preventing, or delaying, the aging of the skin. Through their surfactant action liposomes also improve the coagulation and sinking of oil spreading at a water interface, a methodology which has been under study for some time by EPA for cleaning up oil spills.

Formation of liposome

The phase diagram in Figure 3 illustrates the phase changes around the transition temperature (T_c) of a phospholipid-water system. This phase transition temperature is defined as the minimum temperature required for the water to break through the lipid membrane. When the system is cooled to temperatures below Tc, the hydrocarbon chains adopt an ordered packing phase, thus creating a lamellar structure.

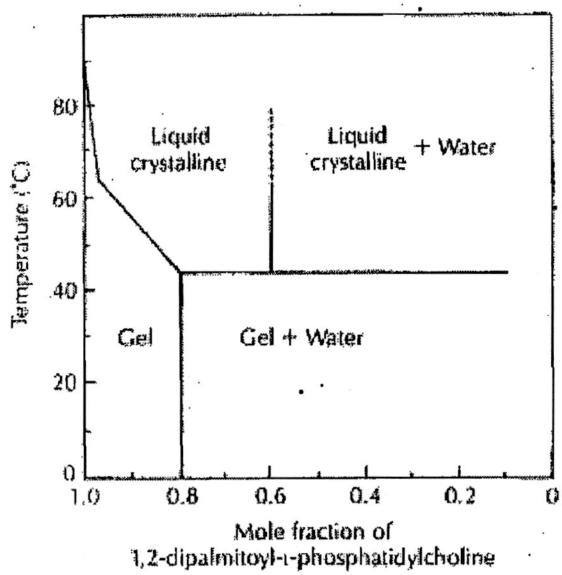

Figure 3. Phase diagram of 1,2-dipalmitoyl-l-phosphatidilcholine water.

In the bilayer structure, the hydrophobic tails are lined up together through hydrophobic interactions, whereas the hydrophilic part of the lipid faces towards the aqueous phase.

Microencapsulation of liposome

The lipid-aqueous system needs to meet two major requirements for the liposome microencapsulation to occure. First, the system needs have a negative free energy, and second, it needs to overcome the energy barrier which is necessary for the formation of the bilayer. Three methods of liposome preparation are here described:

Microfluidisation: Microencapsulation by this method is obtained through the dynamic interaction of two pressurised aqueous-lipid fluids that create a large momentum and flow turbulence that allows the system to overcome the energy barrier to microcapsule formation. The pressure applied in air-driven microfluidisers can be as high as 10,000 psi. The ultra-high velocities reached by this technique allow the creation of small liposomes (< ~0.3μm) with high capture efficiency. This system is useful because of its capability to produce very large amounts of liposomes with adjusted size in a continuous process.

Ultrasonication: The ultrasound absorption is employed to overcome the energy barrier. The sonication of the lipid emulsion can be carried out by two different approaches. The first one is through the use of a sonication probe placed directly into the suspension of liposomes. The second method is slower than the first one and employs a sonication bath, such as a sealed container filled with Nitrogen or Argon gas. Both methods have been extensively applied for the formation of SUV's; however, the use of a sonication probe has been found to cause contamination of the liposomes.

Reverse phase evaporation: Reverse phase evaporation is used for the preparation of LUV's and it is based upon the extraction of a nonpolar solvent from an aqueous-nonpolar inverted micelle by rotatory evaporation under vacuum. This withdrawal of the nonpolar phase changes the intermediate gel-like phase of the micelle into uni-lamellar and oligo-lamellar vesicles. The advantage of this technique is the uniformity of the vesicles formed (from about 0.2 to1.0 μm) as well as their high encapsulation efficiency. On the other hand, the exposure of the components to organic solvents and sonication is likely to result in protein denaturation.

Applications of Lectin

Lectins are proteins, or glycoproteins, that agglutinate erythrocytes of some or all blood groups in vitro. They are an important group of bioactive proteins and glycoproteins found in most organisms, including plants, vertebrates, invertebrates, bacteria and viruses, and have several important applications to the fields of health food and medical biotechnology. Lectins are used as tools in the fields of biochemistry, cell biology and immunology, as well as for diagnostic and therapeutic purposes in cancer research.

Lectin aggregation can be employed on a large-scale basis for the commercial production of biologically active proteins. Lectins have been used in glycoprotein purification, oligosaccharide analysis, as well as in cell-selection processes. Lectins can bind reversibly with free sugars or with sugar residues of polysaccharides, glycoproteins or glycolipids. There has been increasing demand for novel diagnostic and medical cancer therapies that utilise non-traditional sources. Epidemiological studies indicate that the consumption of a plant-based diet is strongly associated with a reduced risk of developing several types of cancer.

Plants contain numerous phytochemicals that can alter cancer-associated biochemical pathways. One such group being intensively examined for its role in cancer chemoprevention is lectins. Lectins are currently being considered for use as cancer therapeutic agents. Lectins were reported to bind preferentially to cancer cell membranes, or their receptors, causing cytotoxicity, apoptosis and/or tumor growth inhibition. Lectins were thought to become internalised into cells, and some lectins were claimed to cause cancer cell agglutination and/or aggregation. Present in common foods, some lectins resist acid and/or enzymatic digestion and also were reported to enter the bloodstream in an intact, and biologically active, form. Lectins possess a spectrum of beneficial, as well as harmful, effects both in vitro and in vivo.

Ingestion of lectins also sequesters the available body pool of polyamines, thereby claimed to thwart cancer cell growth. They have also been reported to affect the immune system by altering the production of various interleukins, or by activating certain protein kinases. Lectins were also reported to bind to ribosomes and thus inhibit protein synthesis. Lectins may also modify the cell cycle by inducing non-apoptotic G1-phase accumulation mechanisms, G2/M phase cell cycle arrest, and apoptosis, and might activate the caspase cascade. Lectins were also reported to down-regulate telomerase activity and inhibit angiogenesis.

Lectins could inhibit cell adhesion, proliferation, colony formation and hemagglutination, and were reported to have cytotoxic effects on human tumor cells. Lectins could function as surface markers for tumor cell recognition, cell adhesion, signal transduction across the membrane, mitogenic cytotoxicity and apoptosis. Also, lectins were reported to modulate the growth, proliferation and apoptosis of premalignant and malignant cells both in vitro and in vivo. Most of these effects are thought to be mediated by specific cell surface receptors. For many years, lectins have been considered toxic substances to both cells and animals, mainly because of the observed agglutination of erythrocytes and other cells in vitro.

On the other hand, it has also been reported that lectins have an inhibitory effect on the growth of tumors. Their potential for clinical applications has been investigated only in recent years. Lectins are now being considered for use both in the diagnostics and therapeutics of cancers. Thus, lectins are quite versatile biomarkers and have been utilised in a variety of studies involving histochemical, biochemical and functional techniques for cancer cell characterisation. Lectins may also be very useful tools for the identification of cancers and the degree of metastasis, or cancer development stage.

Recently, there has been a tendency to shift lectins use from cancer detection to actual use in combating cancer. The reason for this shift is mainly caused by recent research that indicated the cytotoxic and apoptosis/ necrosis-inducing effects of certain lectins, combined with the hypothesis that dietary lectins enter the systemic circulation intact. One important feature appears to be that lectins stimulate the human immune system. Lectins were thus reported to exhibit antitumor and anticarcinogenic activities that could be of substantial benefit in cancer treatment.

Extracts of Viscum album (mistletoe) are widely used as complementary cancer therapies in Europe. Mistletoe has been used parenterally for more than 80 years as an anticancer agent with strong immuno-modulating action. The quality of life of patients with late-stage pancreatic cancer was reported to be improved on account of exposure to mistletoe lectin. Immuno-modulation using recombinant ML was reported to influence tumor growth in breast cancer patients. Bladder carcinoma was reported to be significantly reduced, and survival times were reported to be prolonged in mice as the concentration of ML was increased from 3 to 30 ng. ML increased the life span, decreased the tumor growth and decreased hyperplasia of mice and rats with lymphoma and lung cancer.

A lectin purified from mesquite seed was reported to have an anti-proliferative effect on the cervical human tumor (HeLa) cells and on cell adhesion. Interestingly, mesquite lectin modulated the growth, proliferation and apoptosis of HeLa cells, while having no effect on normal cells in vitro. Vicia faba agglutinin (VFA), a lectin from broad beans was reported to aggregate, stimulate the morphological differentiation of, and reduce the malignant phenotype of colon cancer cells.

Wheat germ agglutinin (WGA) proved to be highly toxic to human pancreatic carcinoma cells in vitro. WGA exposure induced chromatin condensation, nuclear fragmentation and DNA release, consistent with apoptosis. The binding of the snail lectin Helix pomatia agglutinin (HPA), which recognises N-acetylgalactosamine and N-acetylglucosamine sugars, is considered to be a strong predictor of metastasis and unfavourable prognosis in a number of human adenocarcinomas, including breast cancer.

Because of their carbohydrate bio-recognition properties, lectins may also be used as carriers for targeted drug delivery, in a manner similar to liposomes, provided the possible side effects of such treatments could be minimised. It has been observed that mucosal expression of terminal unsubstituted galactose is increased in colon pre-cancerous conditions and cancer, and that it allows interaction with mitogenic galactose-binding lectins of dietary or microbial origin. Based on this observation, an interesting hypothesis was postulated whereby galactose might be able to prevent cancer by binding and inhibiting such lectins from interacting with colon cancer cells.

D-galactose treatment was reported to be effective in liver lectin blocking to prevent hepatic metastases in colorectal carcinoma patients. Epithelial cancer cells showed an increased cell surface expression of mucin antigens with aberrant O-glycosylation, notably Thomsen-Friedenreich Antigens (TFA). TFA is a carbohydrate antigen with a proven link to malignancy. Immunoassays could be utilised for antigens such as TFA in order to determine the metastatic potential of breast and colon cancer cells.

Molecular changes in the membrane surface in the case of both stomach and colon cancer cells occur during the progression to carcinogenesis. Carbohydrate patterns displayed on the cellular membrane exterior are molecular signatures with unique biological characteristics related to oncogenesis and metastasis, and could be used to determine the appropriate chemotherapeutic and surgical procedures for each specific cancer. Lectins have already a demonstrated potential for the treatment, prevention and diagnosis of chronic diseases such as cancer. Further research is, however, required to further elucidate the effects of purified and dietary lectins and their potential for defense against tumors.

Computational Biology

Computational Biology has a very wide range of applications currently thought of 'belonging' to Biotechnology, such as Bioinformatics, even though the development of such computations has preceeded modern Biotechnology by many decades. Instead of attempting the hopeless task of covering superficially just a few of the applications of Computational Biology to either Medical or Agricultural Biotechnology, we decided in favour of an approach that focuses on a few selected examples in greater depth, by considering molecular modelling techniques that have very wide applications, ranging from 'pure' chemistry, to biochemistry, molecular biology, biotechnology, medicine, foods and industrial manufacturing.

Molecular Modelling

Molecular modelling is a group of techniques that employ computer-generated images of chemical structures that show the relative positioning of all the atoms present in the molecule being studied, and/or the simulated dynamics of such molecules together with their ordering through space-time. Such techniques are of considerable help for understanding many physicochemical properties of molecules, and may also provide clues about their possible role(s), that is, their function, in the organism. They can be thus especially valuable tools for investigating structure-function relationships.

Proteins-within a given protein family-have, in theory, similar sequences and generally share the same basic structure. Thus, once the structure for one member of the protein family is determined, molecular modelling computations can help determine the structure for other members of the same protein family. Such a homology technique when applied to protein structure may allow scientists to gain additional insight into protein structure, especially for those proteins for which the available experimental data is scarce.

Structures of apolipoprotein

Lipoprotein in mammals have evolved as the primary transport vehicles for lipids. This role leads to the importance of lipoproteins in several diseases, such as atherosclerosis and cardiovascular disease. Lipoprotein particles consist of a core of neutral lipids, stabilised by a surface monolayer of polar lipids complexed with one or more proteins. Apolipoprotein A-I and apo B are respectively, the major protein components of highdensity (HDL) and very low density (VLDL) lipoproteins. Thus, understanding apolipoproteins is very important for medical and health-related fields, such as medical biotechnology, as well as food science and human nutrition.

The process of biosynthesis, the physical characteristics and the metabolism of apolipoproteins have been intensely studied. However, because of the noncrystalline structure of many apolipoproteins, it has been difficult to obtain structural data at the molecular, or atomic, level. Therefore, methods combining the amino acid sequence with molecular methods are now being introduced. Thus, overall structures may be derived from correlations of global secondary structures determined from polarised light studies combined with local structures predicted from amino acid sequences. The known amino acid sequence of apolipoprotein from the position 7 to 156 of apo Lp-III was first used to design an Apo A-I template, that could be then approached by 'standard' molecular modelling techniques.

Molecular modelling of apolipoproteins A-I

Target: Apolipoprophorin III is chosen as a target for apolipoprotein A-I. (apo Lp A-I). The structure of Apolipophorin III has been determined in a crystal at 2.5Å resolution for the 18-kDa apo Lp III from the African migratory locust, Locusta migratoria and the 22-kDa Nterminal, receptor-binding domain of human apo E.

Template for Apo A-I: Lp IIIa is designed by using molecular software IALIGN from Lp III by inserting alanine for template of sequences of apo A-I (using the programme SYBYL, V5.5). Alanine residues are inserted at each of the gap position identified by IALIGN, an interactive alignment programme distributed with the Protein Identification Resource (PIR). This model was compared with DgA-I, HuA-I and ChA-I resprsenting canine, human and chicken Apo A-I respectively. Results were then compared using a "strip of the helix" template by scoring 1 or 0 for residues that did, or did not, fit into the template. Modelling results:

— *Sequence Comparison*: The five long a-helices connected by short loops in amino acid residues 7-156 of apo

Lp-III is used as template for LpIIIa, Dig A-I, DgA-I, HuA-I. Amphipathic potential (AP) is used to detect if the predict structure is suitable for a lipid-aqueous interface in its stable condition.

— *Energy minimised models*: In this model, electrostatic interactions contributes the most to favourable energies. Alanine was chosen as the spacer residue in building apo Lp-IIIa because of its function of small, non-ionic side chain and serves as a helix-stabilising residue. Although it has a high probability of being found in helical structures, it does not participate through electrostatic interactions. Results of this model are discussed for potential energetic evaluation and amphipathic analysis of energy refined helices.

After evaluating the potential energy for this model, the lateral view structures of apolipoproteins for apo lipophorin III (residues 7-156), canine apolipoptrotein A-I (residues 72-236), human apolipoprotein A-I (residues 73-237) and chicken apolipoprotein A-I (residues 72-236).

Combination of Molecular Modelling with other Techniques

It is important to combine molecular modelling with other techniques in order to improve the accuracy of the modelling results. A recent study of Apolipoprotein has produced a highresolution reconstruction of the structure of apolipoprotein through combination with a solutionphase X-ray technique. It was shown that Apoliproprotein A-1 is an 243-residue protein that contains a globular amino-terminal domain (residue 1-43) and a lipid-binding carboxyl-terminal domain. The aqueous phase X-ray crystal structure was obtained at 4Å resolution. This has suggested the accuracy of molecular modelling by combining the X-ray crystallographic with molecular modelling.

Analysis of Molecular Model

The determination of apolipoprotein A-I can be then further associated with lipid containing domains by employing other molecular modelling techniques. The 'belt model' is used to show the possible orientations of lipoprotein with its apolipoprotein inserted. The suggested structure can then serve as a template in other high density lipoproteins for their structure determination and also help in understanding the biological interaction.

Applications

Molecular modelling has been introduced for more than two decades ago. Increasingly, modelling software is available for a variety of industrial applications. Global markets for molecular modelling, in general, now exceed 2 billion US $ annually.

Applications in the food industry: Molecular modelling has been suggested by several professionals in food industry as a new tool for food research. Such tools can assist food scientists in problem solving, as well as save time and money. Examples of utilisation of molecular modelling are the uses of high intensity sweeteners and taste receptors to predict the sweetening potential of new molecules by using molecular modelling as developed by and E.W. Taylor and S. Wilson at the University of Georgia. Such models can be used in food industry for product development and also for faster results in sensory evaluation.

Medical Applications: Molecular modelling is especially helpful in medical fields, such as in development of new drugs on a nanoscale. Recent studies have shown the importance of using molecular modelling in both medical and food sciences (Food Ingredient first, 2003). The molecular modelling of Epigallocatechin Gallate (EGCG), and the HIV cell was undertaken by Shearer. His report has inspired scientists in Japan who discovered the potential of green tea as an anti-HIV drug. The chemical compound that is found abundantly in the green tea called Epigallocatechin Gallate (EGCG) is reported to stop the HIV virus from binding to CD4 molecules and human Tcells.

Other applications: Other applications of molecular modelling to manufacturing, life sciences and chemistry greatly benefit from such molecular modelling programmes. Nanotechnology has developed to a 30 to 40 million US $ market, and it also has the potential to grow to a 60 to 70 million US $ market within the next five years.

4

Agricultural Applications of Biotechnology

Gene technologies are going to be of great value to agricultural production in a world faced with an ever increasing demand for food. In some instances characteristics of the biological and agricultural systems in which they are to be applied will warrant careful thought and control over the introduction of the new technologies. There could be significant ecological hazards. Also, there may be, particularly in developed countries, ethical or other social issues raised about the use of transgenic food plants.

However, the economic and humanitarian imperatives of their use cannot be ignored. What is critical is that in the implementation and development of the use of transgenic crops, ecological and evolutionary considerations are drawn into the decision making processes at an early stage. This will ensure that appropriate safeguards and management strategies can be devised in parallel with the development and trialing of any transgenic cultivar.

In Asia, grain production will need to double and Latin America faces a major challenge. In Africa, yields are currently well below the genetic potential of the crops that are grown, so efficiencies are imperative but much food will still need to be produced elsewhere. These interactions mean that the demand for food will increase by close to 100% in the first quarter of the 21st century. Where and how is this food going to be generated? The impressive food production increases of the past 30 years have been achieved through the synergistic interaction of better management, increased inputs and increased genetic potential. Is it realistic to expect continuing success at this or an even greater rate, especially in the light of increasing additional constraints arising from mankind's belated recognition of the parlous state of the world's environment?

More than ever before, producers are being forced to take a longer-term view of the environment, increasing yields whilst taking care of the biotic and abiotic natural resources of their lands, ensuring sustainability of the production enterprise. From the producer or farmer's point of view, the objective of his agriculture is to maximise profits whilst providing for a continuing production enterprise. Unfortunately though, given the vagaries of weather conditions that effect crop production, and unpredictable fluctuations in both buyers' preferences and the volume of competing production elsewhere in the world, care for the natural resources of cropping areas often takes a back seat to the immediate economic imperatives of short-term survival.

We need to recognise that sustainability of production systems will not be achieved without reliable profitability being built into the system. Profit comes from both yield levels and, increasingly in different crops, quality differentiation fitting the product to market requirements, this usually attracting a premium price. In both these areas, genetic improvements and management practices contribute markedly to the farmer's livelihood. Genetics really is concerned with yield potential whereas management is important in the longer term for sustainability, and in the shorter term for achieving optimum yields with minimum and timely inputs. Improvements brought about by both genetics and management practices are important for the farmer but management practices alone, at this time, are addressing the sustainability issue, the agricultural system component.

Biotic and Abiotic Resources

Production agriculture operates in modified environments-even in subsistence agriculture natural environments are changed. This need not be detrimental to adjacent natural environments nor need it be debilitating to the production areas, however agriculture has unfortunately led to negative effects on the biotic and abiotic components of the environment. With regard to the care of biotic resources, extensive genetic resource collections for the major cereals exist, and as far as can be judged, a large proportion of the primary and secondary gene pools of these crops is now in safe keeping and available for plant breeding programmes.

Although much work has still to be done to improve the intellectual and physical infrastructure needed for efficient accessing of the genes stored in these gene banks, plant breeders have, and continue to successfully use these resources to develop higher yielding varieties, especially of rice, wheat and maize. Other grain crops such as sorghum and millets are covered by large collections but basic biological knowledge of their representativeness is incomplete. In general, with the exception of potato, the world's major starchy crops have received far less attention than the cereal crops.

One of the difficulties for the starchy crops is that conservation is generally in field gene banks; as a consequence existing collections are primarily stocked by cultivated varieties. However, recent advances in in vitro conservation methods are making it possible to envisage a more extensive set of collections in the future. Knowledge and care of the species interacting with the production species is another matter, particularly of soil microorganisms-knowledge in this field and consequent good practice are in their infancy. Past agricultural practices have often given no thought to the maintenance of all-important abiotic resources.

As a consequence, erosion, increasing acidification and salinisation have become extreme in many parts of the world, taking what was previously some of the best land out of production. Increasing awareness of the despoiling of both the soil and water resource is leading to the development and use of better management practices with a considerable research focus building on water availability and the appropriate use of fertilisers. Genetics will certainly play a role in both countering and reversing these negative trends, yet the genetic improvement of crop species for greater efficiency in water use and in the acquisition of nutrients, is only just becoming a reality.

Agricultural Biotechnology in Developing Countries

The Latin America and Caribbean (LAC) region harbors major centres of origin and diversity for a number of organisms that support current world food production and industry. Agriculture and biodiversity are closely linked to biotechnology; the sustainability of protected biodiversity areas depends to a large extent on what transpires in intervened adjacent areas. Sound agricultural intensification can prevent further deterioration of protected landscapes.

In traditional agricultural areas, modern biotechnology offers an opportunity not only to increase productivity and significantly reduce the use of off-farm chemical inputs, but also to enhance natural agro-biological systems. Biotechnology has now become a preferred tool for adding value to biodiversity by allowing more effective identification and utilisation of genes and derived products. Recent advances in genomic, proteomic and metabolomic research offer unprecedented opportunities for the search, identification, and commercial utilisation of biological products and molecules in the pharmaceutical, nutraceutical, agricultural, and environmental sectors.

This is crucial in a region where the agro-industry sector contributes 25% to the internal gross product (IGP), where the population is forecasted to reach 663 million (about one tenth of the world total) by 2020, and where there is also a growing demand for food quantity and quality, for preventive control of human diseases, for mitigation of environmental degradation, and for increasing income by tackling new markets competitively. Some studies on the status of plant biotechnology in the LAC region have been carried out.

Early attention given to tissue culture research allowed the establishment of in-vitro germplasm banks for several important crop plants in the LAC countries. Cryoconservation was developed to some extent for potato and cassava by the International Potato Center (CIP) in Peru and the International Center for Tropical Agriculture (CIAT) in Colombia, respectively, and for coffee somatic embryos by the Tropical Agronomic Research and Teaching Center (CATIE) in Costa Rica. Genetic diversity analysis, using molecular markers, has improved the characterisation and conservation of economically important crop germplasm, and recently has been extended to promising Amazonian and Andean species.

Today, advanced biotechnologies (such as DNA microarrays) are beginning to impact genetic resource assessment and utilisation research; however, because of the high investment and expertise required, these developments are limited to certain groups. They are currently being applied mainly in Brazil, but Mexico, Cuba, Venezuela, Chile, and Colombia are also developing genome research. Andean root and tuber crop biodiversity studies are underway in Peru.

Applications to Plant Health

Monoclonal antibodies, recombinant antigens, and molecular plant disease tests have greatly improved diagnostic systems and pathogen characterisation. The most widely used technique for generating pathogen/ virus free plants in the region involved the combination of thermotherapy and meristem tip culture. This tissueculture-based application is still actively being used in most seed-production programmes of the region. The CIP and CIAT pioneered the development of these techniques for generating and distributing healthy plant stocks of vegetatively propagated crops (e.g., potato, sweetpotato, and cassava).

National programmes and private groups of Brazil, Chile, Uruguay, Cuba, Mexico, Argentina, Colombia, and Peru have developed and commercialised these technologies, to variable extent, for the production of healthy seed. Tissue-culture-based propagation techniques have been used in the region over many years for the multiplication of planting material of many economically important crop plants, such as roots and tubers, flowers, fruit trees, and others. These propagation techniques are now being standardised for a range of native plants of commercial value, such as forests and ornamentals.

Bioreactors and temporary immersion systems are now available for industrial plant multiplication; however, development of this technology is still incipient in the region. Important applications have been demonstrated for coffee somatic embryo multiplication at CATIE, for cassava multiplication, rice anther culture, and regeneration of genetically transformed cassava at CIAT, and for the regeneration and multiplication of potato clones at CIP.

Genetic Improvement

For many LAC biotechnology groups, anther culture for haploid plant production and embryo rescue for hybrid plant production have been connected to crop improvement programmes, such as those in Mexico, Colombia, Costa Rica, Uruguay, Chile, and Brazil for rice and grain legumes. Molecular marker-assisted selection has improved plant breeding through the use of tightly linked markers in assisted selection. Practical examples exist in the region for maize and wheat at CIMMYT; cassava, rice, and common bean at CIAT; potato and sweetpotato at CIP; and in the national programmes of Brazil, Mexico, Uruguay, Chile, Colombia, and Costa Rica. The LAC region has conducted 27% of the world's field trials of 24 transgenic crops.

Argentina alone represents 21% and is second in the world in growing transgenic crops commercially; this has represented \$200 million of profits to Argentina in 2000. Argentina, Brazil, Colombia, Honduras, Mexico, and Uruguay commercialise genetically engineered (GE) crops. Brazil officially approved GE crops for planting in 2003 and has already become the fourth in the world in growing GE crops. Colombia approved 10,000 hectares of commercial Bt cotton and carnations for export. Bolivia has carried out field trials for GE cotton and soybeans, and Honduras is growing GE maize for the first time.

In Mexico, the Savia firm has performed within the ten top worldwide. Because the global market for transgenic crops is projected to reach $25 billion in 2010, this application area has generated large expectations in LAC countries. Biosafety and IPR regulations still have to be enforced in many countries for an effective and safe use of genetically engineered crops, especially if their production is meant for the export market. The agricultural systems that prevail in significant sectors of the LAC region are characterised by a mosaic of continuous cropping systems, complex crop/pest management systems, and biological, cultural, and socioeconomic diversity, all of which need to be considered in the adoption strategies of any agricultural new technology, including genetically engineered crops.

Currently, the most commercialised biotechnological product in the LAC region has been transgenic seed, but it has been concentrated basically in one country. The next in importance is the selling of virus-free stocks and seeds, and in third place are biopesticides and other agricultural bioinputs. Countries which have used modern biotechnologies like genetic engineering have had a higher value of their seed market than those using mature technologies like tissue culture.

The FAO's 2000 report pointed out to a general weakness in the quantity and quality of regional biotechnology researchers: Only 40% of the limited number available were postgraduates, and only about 10% had doctoral degrees. Most researchers were biologists or agronomists, with very few specialised in key areas such as molecular genetics, molecular biology, protein engineering, molecular and industrial microbiology, or bioinformatics. An aggressive programme of personnel development needs to be implemented in most LAC countries in basic sciences and cutting-edge biotechnological applications in particular, in the Central American and the Andean regions-otherwise any biotechnological programme aimed at generating income and employment would face serious drawbacks.

By 1999, the LAC region invested only 0.59% of the IGP in research and development, with Brazil, Chile, and Cuba showing the highest level. However, even these lagged behind countries like Canada (1.61%), France (2.18%), the United States (2.84%), and Japan (3.06%). Dellacha et al. reported that 62% of the scientific and technological activities in LAC were funded through the state's national science and technology councils, 28% by companies, and 9% by universities; funding coming from external sources was approximately 1%. State support and private investment for biotechnology research varies in the region. In some countries, it is minimal; for example, in Peru the 1998 funds from the state amounted only to 0.06% of the IGP, while in Brazil state funding was 0.86% in 1999.

Biodiversity-based Biotechnology in Developing Countries

The LAC region concentrates major biodiversity hotspots of the world. The region is also a center of origin and diversity of a number of species that sustain current world food supply (e.g., potato, sweetpotato, corn, tomato, beans, cassava, peanuts, pineapple, cacao, chili pepper, and papaya). Furthermore, the greatest number of flowering plants with unusual sources of compounds for food and agriculture, as well as for the biopharmaceutical, nutraceutical, cosmetic, and environmental industries, exists in this region. This comparative advantage needs to be explored and sustainably utilised by integrating the emerging biotechnologies with the region's rich traditional knowledge on the properties and attributes of its biological resources.

The world market for biological resource-derived ingredients and molecules in 1999 was US$925 billion for different sectors and industries. Several LAC research groups have begun to work with native forest and medicinal species with the view to explore commercial opportunities in the use of natural ingredients. Application of simple bioprocessing technologies offers a short-term approach for the production of sweeteners, flavour products, fruit juices, amino acids, pigments, vitamins, and antioxidants. For example, the Agroindustrial Group Backus S.A. in Peru has used camu-camu, an Amazonian fruit of the Myrtaceae family, for ascorbic acid extraction.

The nutraceutical and biopharmaceutical products have become the third and fourth most important markets for biological resource-derived products. In order to tackle these emergent markets, new scientific specialisations are being created in universities and industry worldwide. Some have resulted from the merging of former branches of science; for example, ethno-botany has joined with pharmacology to form ethnopharmacology. On the other hand, new tools are being developed for the automatisation and miniaturisation of genetic and chemical analyses and syntheses (e.g., robotics, DNA chips, combinatorial chemistry, liquid chromatography, mass spectrometry, X-ray spectrophotometry, nuclear magnetic resonance, among others).

The so-called new "omic" sciences (functional genomics, proteomics, and metabolomics) and bioinformatics are improving the search of chemical principles from plant species to develop new target-specific drugs. Not only are specialised personnel and modern equipment needed to achieve scale-up capabilities in this field, but appropriate business management, including IPR, are major factors that still limit commercial takeoff in the LAC region.

Since 1997, Brazil has implemented several genome projects in universities and private research laboratories under the management of a virtual institute, the Organisation for Nucleotide Sequencing and Analysis. The organisms selected include bacteria and plants important to Brazil's economy and also of global interest (e.g., sugarcane, Xylella fastidiosa, Xanthomonas species, and Eucalyptus). These studies involve high-throughput sequencing, functional genomics, proteomics, and transcriptomics, among other advanced technologies. The initial goal was the sequencing of the plant pathogen Xylella fastidiosa, which was achieved for the first time worldwide.

Other efforts in the region are in Mexico, Cuba, and recently in Chile, where the national genome project will tackle the functional genomics of grapevine and peach. A Latin American Genome Biology Network, sponsored by the United Nations University for Latin America and the Caribbean (UNUBIOLAC), has been created with the objective of enhancing linkages among Latin American groups working with high-throughput genomic technologies.

In 2003, a study in Andean countries (Bolivia, Colombia, Ecuador, Peru, and Venezuela) supported by CAF (Corporación Andina de Fomento) and CEPAL (Economic Commission for Latin America and the Caribbean) was initiated with the goal of analysing biotechnological, institutional, and human-resource capacities involved in biotechnology research and development for biodiversity utilisation and generation of biotechnological products and biotrades. The study also analyses the global/regional market trends for their potential commercialisation and eventually will offer recommendations and strategic guidelines in support of these countries to promote sustainable value-added transformation and commercialisation of products derived from biodiversity and biotechnology.

The regional CGIAR centres, the CIP, CIAT, and CIMMYT, and the regional organisation (CATIE) have played active roles for many years in the regional development of biotechnology applications to selected crop plants through the diffusion of advanced strategic technologies (including genomic), have offered technical and scientific training, and have developed collaborative work with national research institutions. REDBIO has had, and continues to have, a critical role in developing biotechnology science and applications in the LAC region.

We believe it is now time to turn attention to biodiversity-based biotechnology research and development. Another promising future role of biotechnology such as nitrogen fixation, photosynthetic efficiency, soil nutrient cycling efficiency, abiotic (drought, acidity, salinity) stress adaptation, and the reduction of off-farm inputs such as agrochemicals. All have far-reaching implications for sustainable agricultural systems in LAC agroecologies like the Andean region. At present, however, most commercial biological products used in countries like Colombia and Cuba have been elaborated using the bacterium B. thuringiensis, the fungi Beauveria, Metarhizium, and Paecilomyces, and baculoviruses.

In recent years, private investment in agricultural R&D has increased, particularly in countries where effective and transparent regulatory frameworks are in place and comparative research infrastructure and qualified human resources exist. A dynamic biotechnological industry, with clusters of companies and public research institutions, has begun to emerge in some LAC countries. Joint public/private ventures have explored collaborations to address bioproduct market demands. One example is a bioprospecting collaboration, under an ICBG project, between the Aguaruna community of Peru, three universities, and the G.D.

Searle Corporate Partnership, to apply ethnomedicinal approaches for examining plant biodiversity based on the Aguaruna pharmacopeia and involving a wide range of human diseases and syndromes. Patent applications have been filed as a result; benefit sharing with the communities has been included. Another ICBG project in Panama is searching new bioactive compounds against cancer and the tropical parasites causing malaria, leishmaniasis, and Chagas' disease, through a collaboration between Smithsonian Tropical Research Institute, several universities in Panama and one in the United States, and Novartis.

In Cuba, the Biotechnological West Pole of La Havana comprises more than 40 specialised biotechnology centres; others include the BioTrade initiative for the promotion of biodiversity-derived products and services in Colombia, under the leadership of the Alexander von Humboldt Institute, and the emerging companies in Amazonia supported by UNCTAD-Bolsa Amazonica-IavH. The adoption of intellectual property and biosafety regulations has recently been promoted, but management and enforcement varies among LAC countries.

Chile is the only country where biotechnological processes can be patented. Microorganisms can be patented in Brazil and Mexico, but neither microorganisms nor genes can be patented in the Andean countries. In most countries, UPOV-type plant variety protection systems exist. Stimulation of investments and facilitation of the acquisition of technologies through collaborative partnerships should go hand-in-hand with mechanisms to link the research with the holders of biological resources. Governments can offer tax and other incentives to investors; these incentives should encourage the sharing of the derived benefits with the research partners and with the traditional curators of genetic resources.

Finally, it is relevant to mention that food insecurity in large sectors of the LAC region is a consequence of inadequate social, economic, and technological development. The new biotechnologies open a range of opportunities for increasing biodiversity-based product diversification. It is necessary to bear in mind not only the local socioeconomic context, but also the level of export/import of agricultural and biodiversity derived products; the importance of the small, medium, and large agro-industry in the economy; the country's research and technological capacities; and the existence of a legal framework that stimulates biodiversity conservation and utilisation. The situation of the resource-poor farmer should be taken into account, to make sure that the benefits of modern biotechnologies reach this important sector-currently a majority in some LAC countries.

5

Biotechnology in Livestock Production

Biotechnology will improve the life of every and allow more sustainable living. The use of biotechnology will lead to a distinct shift in the economic returns from livestock. Livestock is becoming increasingly important in the growth of agriculture in developing economies. The contributions made by livestock to both agriculture and gross domestic product (GDP) have risen, at a time when the contribution of agriculture to GDP has fallen. The demand for livestock products is a function of income, and sustained growth in per capita income, rising urban populations and changes in diet and lifestyle are fuelling growth in livestock production. Livestock production contributes to socioeconomic development in many ways, by augmenting income and employment and reducing the incidence of rural poverty.

Though the role of livestock in ensuring nutritional security is recognised in mixed crop-livestock systems, the importance of livestock goes beyond direct food production. Livestock supply draught power and organic manure to the crop sector, and hides, skins, bones, blood and fibre are used in many industries. Thus, livestock are an important source of income and employment, helping to alleviate poverty and smooth the income distribution among small landholders and the landless, who constitute the bulk of the rural population and the majority of livestock owners. In addition, livestock can easily be converted into cash and thus act as a cushion against crop failure, particularly in less favoured environments.

By enabling crop residues and by-products to be used as fodder, livestock production contributes positively to the environment. Animal owners in the developing world are predominantly resource-poor small-scale operators with little or no land and few animals, who must operate within the constraints of the local climate and who have limited purchasing power and little access to resources or opportunity to determine resource allocation for animal production. The situation of the poorest livestock owners is fast deteriorating owing to the fragmentation of limited holdings, exhaustion of land resources and increasing human and animal population pressure. Low livestock productivity in many developing countries is considered to reflect, among other things, the inadequate supply of animal husbandry and veterinary services. Veterinary services have traditionally been provided by the State, but financial constraints have limited the availability and effectiveness of public services.

Benefit of Technology

A major benefit of agricultural research and technology is that the purchasing power of the poor increases, because both average incomes and access to staple food products are improved. Studies by economists have provided empirical support for the proposition that growth in the livestock sector affects the whole economy. Rapid growth of livestock production has stimulated demand for and increased the value of land, labour and nonagricultural goods and services, thereby leading to overall economic growth. The poor spend a relatively high proportion of any additional income on food, so increases in livestock production achieved through the use of biotechnology can have major nutritional implications, particularly if the technology is aimed at the poorest producers.

However, studies have revealed that the commercialisation of agriculture has reduced the nutritional security of the poor. Once production of milk, meat or eggs has been enhanced through the use of technology, it is hoped that it will also make a significant difference in other areas such as nutrition, prevention of diseases, healthcare and other management practices. It is in these areas that biotechnology shows promise and is currently being used. Green Revolution technologies (i.e. those technologies designed to improve the efficiency of agricultural processes and increase crop productivity by relying on the extensive use of chemical fertilisers/ pesticides and heavy machinery) are intended to be used in package form (e.g. new plant varieties supplied with recommendations on fertiliser, pesticide and herbicide rates and water control measures); however, among livestock producers many components of these technologies have been taken up in a piecemeal, often stepwise, manner.

The sequence of adoption is determined by availability and by the potential cost savings. The sequential adoptions of crop management technologies for rice and wheat have been assessed in detail, but few similar studies have focussed on livestock production in developing countries. Evidence from the People's Republic of China, Mexico, South Africa and India suggests that small farmers have had no more difficulty than larger farmers in adopting the new technologies. The question, therefore, is not whether biotechnology can benefit small-scale resource-poor farmers, but rather how biotechnology can address the agricultural problems faced by farmers in developing countries.

Biotechnology is a promising new tool in the development of applied agricultural technologies. The challenge is to focus this potential on the problems experienced by developing countries. The introduction of multi-enterprise systems or, more broadly, agricultural diversification is seen as the way forward for agriculture in the developing world; such systems could lift small-scale and marginal farmers out of poverty. For example, rotating rice and wheat cropping with dairy farming yields higher profits.

Introducing multi-enterprise systems involving livestock enhances the purchasing power of farmers and helps them to obtain nutritional security. It also generates rural (both farming and non-farming) employment, thus preventing excessive migration to urban areas, which is a common problem in developing economies. Multi-enterprise systems also support the natural environment and contribute to capital formation, thus leading to higher overall growth in the agricultural economy. The technology, infrastructure and institutions now exist to make the application of biotechnology in the context of a multi-enterprise system involving livestock production economically viable.

Advantages of Livestock Production

The multiplicity of genes, species, populations and agroecosystems in the developing countries of South and South East Asia, the People's Republic of China, Africa and Latin America is viewed as a valuable resource for the genetic improvement of livestock on a global scale. The livestock in these countries, which are an integral part of a fragile ecosystem, are a rich source of animal biodiversity. Buffaloes, sheep, goats, camels and zebu cattle have adapted to their regional environments over thousands of years and have provided an important source of sustenance for the population of the region. Livestock production in the developing world has a number of advantages over production in more developed countries, for example:

— the unique and valuable production traits of buffaloes, cattle, sheep, goats and camels

— the low-input production system

— the low unit cost of production

— the lean meat produced from sheep, goats and buffaloes

— the considerable biodiversity

— animal breeds that are resistant to stress and to particular diseases

— the ability of the animals to survive on high-roughage feeds

— the potential for biopharmaceutical developments to lead to significant benefits

— the potential for expanding the microbial food, feed and leather industries

— the integrated production system tailored to the local ecology

— the potential for integrating knowledge and industry.

Several genes and desirable traits have been identified in the livestock of developing countries in Asia and Africa, and some of the livestock species and breeds from these countries have become major contributors to the economy of South America. Examples of breeds from the developing world that are particularly important on a global level are:

- buffaloes that produce milk with a high fat content or with the protein quality required to produce mozzarella cheese
- goats from cold dry regions that produce pashmina and toos (the finest wool in the world)
- Black Bengal goats that carry a gene for high prolificacy
- Garole sheep that carry genes for twinning
- Andaman goats that are highly tolerant of salt
- the yak and mithun that are adapted to high altitude
- the camels, sheep and goats that are adapted to a tropical arid environment and can tolerate feed with a high lignin content
- the many species that are resistant to stress or to particular diseases.

Technological Development in Economy

The genetic resources possessed by animals in developing countries often affect economic development. Animal biotechnology is the result of a multistage process, involving research, development, testing and registration, production and marketing. The goal is to develop a technology, process or product that has clear commercial potential and can be commercialised after due testing and regulatory approval.

Developing countries find it difficult to develop biotechnology because the facilities or resources needed to complete all of the stages in the process are often lacking. However, several technologies from developed countries have been successfully adopted by developing countries. The impact of technology can be analysed by estimating the growth of total factor productivity (TFP) in livestock production. Not many TFP studies on livestock have been reported. However, separate TFP estimates for the aggregate crop and livestock sectors have been made.

TFP analysis has shown a shift towards larger, more commercial and more intensive production systems and has further revealed that, as specialisation has developed over the past decade, the importance of backyard livestock production has declined and the importance of specialised household and commercial enterprise has increased. Studies from India have shown that technological input is responsible for about 45% of total output growth and that the TFP growth may be as much as 1.8%.

There are a large number of technologies that have been developed for or adapted to the livestock of both developed and developing countries. However, the major technologies that are used effectively in livestock production in the developing world include conserving animal genetic resources, augmenting reproduction, embryo transfer (ET) and related technologies, diagnosing disease and controlling and improving nutrient availability.

TRANSGENICS

Although gene-based technologies have the potential to improve the efficiency of livestock production, thereby ensuring better returns for the farmers, the economic impact of transgenics in the livestock sector will be much less than in the crop sector. However, the global adoption of genetically modified (GM) crops, which were grown on 67.7 million hectares in 2003 compared with 2.8 million hectares in 1996, has had a substantial impact on livestock feed. It is estimated that the United States of America (USA), Argentina, Canada, Brazil and the People's Republic of China have 63%, 21%, 6%, 4% and 4%, respectively, of the global transgenic acreage and that the most frequently grown crops are GM soybean (61%), maize (23%), cotton (11%) and canola (5%).

Although few developing countries have released GM crop varieties, a preliminary analysis reveals that more than 20 developing countries are conducting research into the applications of GM crops. Although transgenic animals (especially mice) are used routinely in research (particularly in the medical field), no GM animals have yet been released on farms. A wide range of traits of potential interest to livestock producers have, however, been the subject of research; for example, the gene responsible for the production of growth hormone (which could be manipulated to increase growth rates), the phytase gene (which could reduce phosphorous emissions from pigs) and keratin genes (which could improve the wool of sheep).

The genetic modification of livestock has proceeded much more slowly than the genetic modification of crops for a variety of reasons, including the high costs, the inefficiency of the gene transfer techniques and the low reproductive rates of animals. Recombinant deoxyribonucleic acid (DNA) approaches have been used to promote the expression of desirable genes, to hinder the expression of undesirable genes, to alter specific genes and to inactivate genes so as to block specific pathways. It is estimated that at least 30 enzymes produced by GM bacteria, yeasts and moulds are currently commercially available worldwide; many of these enzymes are used in the food industry. Genetic engineering has been used to introduce foreign genes into the animal genome or, alternatively, to knock out selected genes.

Genes controlling growth were introduced into pigs to increase growth and improve carcass quality. Currently, research is underway to engineer resistance to diseases that affect the animals or that pose an indirect risk to human health, such as Marek's disease and salmonellosis in poultry, scrapie in sheep and mastitis in cattle. Other studies have tried to increase the casein content of milk or to engineer animals that produce pharmaceutical or industrial chemicals in their milk or semen. No agricultural applications have yet proved commercially successful. Nuclear transfer (NT) technology now provides an alternative route for cell-based transgenesis in domestic species, offering new opportunities for genetic modification. Livestock that produce human therapeutic proteins in their milk, that have organs suitable for xenotransplantation and that are resistant to diseases such as spongiform encephalopathies have been produced by NT from engineered cultured somatic cells.

Genetic Diversity in the Livestock

There is considerable genetic diversity in the livestock of developing countries, much of which controls traits that influence adaptability to harsh environments, productivity and susceptibility to disease and parasitism. However, little if any data on these genetic resources are available. Economic analysis can play an important role in ensuring that conservation efforts are appropriately focused. The primary challenge facing conservationists is to identify sound reasons why society should preserve animals that livestock keepers have abandoned.

The breeding practices and breed preferences of livestock keepers can successfully be determined by using research techniques such as the revealed preference hedonic approach. On the one hand, conservation cannot be achieved through a conventional breeding programme because the animals carrying the most advantageous traits cannot be easily identified; on the other hand, conservation cannot be achieved through biotechnology because the necessary technologies are either unavailable or uneconomic. In livestock populations with a high degree of genetic variation, molecular markers are being increasingly used to study the distribution and patterns of genetic diversity.

Global surveys indicate that 40% of domestic livestock breeds are at risk of extinction. Most of these breeds are found only in developing countries, and often little is known about them or their potential. Rapid progress is being made in the preparation of dense microsatellite linkage maps to assist in the search for genetic traits of economic importance. These linkage maps can be used to develop strategies for marker assisted selection and marker assisted introgression that will meet the goals of breeding programmes in developing countries. Molecular markers have been widely used in the identification of genotypes and the 'genetic fingerprinting' of organisms. Genotype verification is used intensively to determine the parentage of domestic animals and to trace livestock products in the food chain back to the farm and animal of origin.

Reproductive Biotechnologies

The main objectives of using reproductive biotechnologies in livestock are to increase production, reproductive efficiency and rates of genetic improvement. Over the years, many options have become available for managing the reproduction of the major large and small ruminants. Artificial insemination (AI) and preservation of semen are the main technologies that are used extensively. Assessing the fertilisation capacity of sperms, sexing sperms, synchronisation and fixed-time insemination, superovulation, embryo transfer (ET) and in vitro embryo production (IVEP) are additional techniques that can improve reproductive efficiency and pregnancy rates. Reproductive technologies can also be used to control reproductive diseases if procedures and protocols are accurately followed.

Artificial Insemination (AI)

The conception rate in field AI programmes in developing countries is very low, and therefore the desired effect in terms of animal improvement has not been achieved. Most semen banks still evaluate semen on the basis of sperm motility, even though significant advances have been made in techniques for semen evaluation. Although detailed guidelines are available regarding the processing, storage and thawing of cattle semen and buffalo semen, the processing and handling procedures in laboratories processing semen are often inadequate. Only when farmers have access to considerably better technical and organisational facilities will AI become more effective.

At present, the efficiency of the technology is limited by organisational and logistical constraints and by the failure to provide appropriate training for farmers. Several modifications of the technique have been suggested to increase the conception rate. Synchronisation with different compounds, and the use of gonadotropin-releasing hormone (GnRH) followed seven days later by prostaglandin $F2_{\alpha}$ ($PGF2_{\alpha}$) can synchronise oestrus and improves the conception rate. In this protocol giving injections of GnRH on day 0, $PGF2_{\alpha}$ on day 7 and GnRH on day 9 is called the 'Ovsynch' programme and synchronises ovulation, permitting timed insemination. The ability to control ovarian follicular and corpus luteum development has allowed insemination in cattle to be timed without having to detect oestrus, and this has increased the net revenue per cow.

Embryo Transfer (ET)

One of the major reproductive technologies that can facilitate genetic improvement in cattle is ET. Unfortunately, commercial ET programmes are limited by the high variability in the ovarian follicular response to gonadotropin stimulation. Multiple ovulation and embryo transfer (MOET) takes AI one step further, in terms of both the possible genetic gains and the level of technical expertise and organisation required. In 2001, 450,000 embryos were transferred globally, mainly in dairy cattle, with 62% being transferred in North America and Europe, 16% in South America and 11% in Asia.

The main potential advantage of MOET for developing countries is that the elite females of local breeds can be identified, and bulls can be produced from them for use in a field programme of breed improvement. Zebu cattle and buffaloes in developing countries exhibit less consistent follicular dynamics after superovulation than Bos taurus in the developed world. However, over the last 10 to 15 years, the number of transferable embryos produced by zebu donors has increased from 2.4 to 5.8 embryos per flush in the late 1980s to 5.6 to 9.9 embryos per flush in 2000.

The use of ET has been less successful than envisaged for several reasons. The low reproductive efficiency, poor superovulatory responses, very low primordial follicle population and high incidence of atresia all contribute to low embryo production. In buffaloes, embryo recovery was initially less than one, but has subsequently improved to 2.6 with 1.4 transferable embryos per flush. After transferring buffalo embryos to recipients, the conception rate is only 16%. The poor success rates have limited the use of ET in buffaloes, which are the main dairy animals in developing countries in Asia, South-East Asia and the Mediterranean region.

In Vitro Embryo Production (IVEP)

Since the birth of the first buffalo calf from an in vitro fertilised oocyte, a number of publications have described the effects of different protocols and media on oocyte and embryo development. Two extensive reviews have been published recently. However, the practical use of IVEP is limited by high production costs and the low overall efficiency under field conditions. The efficiency of blastocyst production in buffaloes is much poorer than the 30% to 60% reported for cattle. Although viable buffalo blastocysts have been produced from ovaries obtained from abattoirs, the yield of transferable embryos remains low (15% to 39%). Embryos produced in vitro have led successfully to pregnancy and calf birth in buffalo, but the success rate is low. Therefore IVEP must be improved before it can be widely used in cattle and buffaloes in developing countries.

Animal Health Care and Biotechnology

Many virulent animal diseases, especially in livestock, are very contagious and can be economically devastating, especially in developing tropical countries. Livestock are susceptible to many types of dysentery and diseases such as African horse sickness, bovine leukosis, bovine infectious rhinotracheitis, brucellosis, and Rift Valley fever, to name just a few. Developing countries often have no way of coping with the widespread outbreak of disease.

Most biotechnologies related to health focus on the needs of the developed world, meaning that 90% of health research is devoted to the health problems of 10% of the world's population. Two main approaches are being used to develop vaccines using recombinant DNA technology. The first involves deleting genes that determine the virulence of the pathogen, thus producing attenuated organisms (non-pathogens) that can be used as live vaccines. Currently, this strategy is more effective against viral and bacterial diseases than against parasites.

Attenuated live vaccines have been developed against the herpes viruses that cause pseudorabies in pigs and infectious bovine rhinotracheitis in cattle. A number of candidate Salmonella vaccines have also been produced. The second approach is to identify protein subunits of pathogens that can stimulate immunity. The International Livestock Research Institute (ILRI) used this approach to develop a vaccine against Theileria parva, the parasite that causes East Coast fever in African cattle. A novel strategy for developing vaccines against bloodsucking parasites involves using components of the gut wall of the parasite that are not usually exposed to the immune system of the host.

When the parasite feeds, it ingests antibodies induced by the vaccine, which destroy the gut wall and, consequently, kill the parasite. This strategy has been used successfully to develop a vaccine against the one-host tick Boophilus microplus. Vaccination is one of the most effective and sustainable methods of controlling disease. Vaccines against parasitic diseases in Africa and viral diseases in Asia have been shown to control disease effectively and increase livestock productivity. A recent approach has been to use vaccines based on DNA.

The use of DNA in vaccines is based on the discovery that injecting genes in the form of plasmid DNA can stimulate an immune response to the respective gene products. This immune response is a result of the genes being taken up and expressed by cells in the animal after injection. The live-vector and DNA vaccination systems could be manipulated further to enhance the immunity conferred by the gene products.

Experimental studies have demonstrated that these vaccines can potentially induce appropriate and enduring immune responses. This technology is, in principle, one of the simplest and yet most versatile methods of inducing both humeral and cellular immune responses, as well as protecting against a variety of infectious agents. However, although immune responses have been induced in a number of larger species, most of the information on the efficacy of DNA immunisation comes from studies of mice. An exhaustive review of the information available on the use of DNA vaccines in farm animals, including cattle, pigs and poultry, has identified the areas that need specific attention before this technology can be used routinely.

These areas include the delivery, safety and compatibility of plasmids in multivalent vaccines and the potential for using immune stimulants as part of a DNA vaccine. Korean scientists have developed a combined vaccine against pleuropneumonia, pneumonic pasteurellosis and enzootic pneumonia in swine. Molecular biology has been used to produce an improved vaccine against swine fever. In the Philippines, a vaccine has been developed that protects cattle and water buffalo against haemorrhagic septicaemia, which is the leading cause of death in these animals. The new vaccine provides improved protection at a very low cost.

Advanced diagnostic tests that use biotechnology enable the agents causing disease to be identified and the impact of disease control programmes to be monitored more precisely than was previously possible. Molecular epidemiology characterises pathogens (viruses, bacteria, parasites and fungi) by nucleotide sequencing, enabling their origins to be traced. This is particularly important for epidemic diseases, in which pinpointing the source of the infection can significantly improve disease control. For example, the molecular analysis of rinderpest viruses has been vital in determining the lineages circulating in the world and instrumental in aiding the Global Rinderpest Eradication Programme.

Enzyme-linked immunosorbent assays have become the standard means of diagnosing and monitoring many animal and fish diseases worldwide, and the PCR technique is especially useful in diagnosing livestock disease. Many diagnostic techniques currently used in developing countries are cumbersome and unsuitable for low-resource settings. Molecular diagnostic technologies that are either already in use or being tested in low-income regions include polymerase chain reaction (PCR), monoclonal antibodies and recombinant antigens. These technologies can be modified to facilitate their application in the developing world.

Simple hand-held devices that rely on the binding specificity of monoclonal antibodies or recombinant antigens to diagnose infection may be easily adapted for use in settings without running water, refrigeration or electricity. Molecular characterisation of the virus serotypes causing foot and mouth disease has helped in the vaccination and control programmes in Asia. In Japan and Taiwan, DNA testing is being used to diagnose hereditary weaknesses of livestock. One test looks for the presence of the gene responsible for porcine stress syndrome in pigs. Pigs with this gene tend to produce pale poor-quality meat because of their reaction to the stress of transport and slaughter.

Pigs with this gene can now be excluded from breeding programmes, so the gene will become less common. In addition, DNA testing is being used in Japan to check for the gene that causes leucocyte adhesion deficiency in Holstein cattle. Cattle with this condition suffer from gum disease, tooth loss and stunted growth.

They usually die before they are one year old. By using DNA testing, carriers can be identified and eliminated from breeding herds. Bulls used for breeding can also be tested to make sure that they are not carriers. Another DNA test identifies a gene that leads to anaemia and retarded growth in Japanese Black cattle.

Detection of Pathogen

New diagnostic tests will increase the sensitivity with which parasites can be detected in animals. Serological tests of livestock populations have proven inadequate for detecting exposure to infectious disease causing organisms. The presence of antibodies does not always indicate current infection, since antibodies can remain in an animal for long periods of time. Instead of detecting the antibodies produced in response to an infection, new diagnostic methods are needed for detecting parasite proteins or antigens and DNA.

Parasite antigen detection kits are available for detecting trypanosome proteins and tick-borne protozoan parasites. Mouse monoclonal antibodies raised against specific parasite proteins are used to identify parasite antigens in animals. These assays are highly sensitive and specific. Often, specific strains or species of an infectious organism produce different responses in an infected animal.

The infecting strain determines which treatments are used. Not only can these assays detect current infections, they can also distinguish between different strains or species of parasite (for example, between the major tsetse-transmitted species of trypanosomes). Parasite-specific DNA assays are desirable because of their sensitivity and because the DNA is stable. Samples can be collected in the field and delivered to a laboratory for DNA analysis.

Polymorphic DNA sequences and their probes allow different parasites, such as tickbome parasites and trypanosomes, to be detected and distinguished. These sequences are detected in both animals and vectors by hybridisation or the polymerase chain reaction. These methods allow animal and vector populations (such as ticks) to be monitored to determine the extent of parasite infection.

Animal Genome

The availability of animal genome sequences using high throughput sequencing technologies will provide a battery of new tools to study disease. Genomic research will provide new insight into pathogenesis, evolution, diagnostics, vaccines, epidemiology, and therapeutics. A well developed bioinformatics approach to genome sequence analysis will yield valuable information for molecular epidemiology of disease outbreaks, for identification of virulence and host range factors and for development of new diagnostics and control approaches.

Current Methods of Treatment

Examples of current methods of treatment for a few important animal diseases are presented below. In many cases, biotechnology is only beginning to offer effective treatments. Developing countries in particular have special needs because of climate (often high temperatures), lack of modern facilities (e.g. refrigeration), and a need for low-cost treatments.

Foot-and-Mouth Disease: This highly contagious disease is one of the most devastating to livestock and very costly to the industry. Worldwide, approximately 30% of cloven-hoofed animals are affected. Although only about 5% of adult animals die from the disease, young cattle and pigs are most affected and approximately 50% of those afflicted die from myocarditis. Infected females abort their calves and rarely produce young. Usually infected animals are killed to avoid spreading the disease and their milk and meat cannot be used.

The most common way of protecting animals against foot-and-mouth disease is by vaccination with inactivated virus vaccine (when available). However, thermal instability, short-term protection (four to six months), and virus strain specificity present problems for treatment and prevention with vaccines. Experimental vaccines include a single inactivated antigen and a mixture of two strains of inactivated antigen for crossprotective activity to different viral strains.

Coccidiosis: Nine parasitic protozoan species in the genus Eimeria invade epithelial cells of the digestive tract and associated glands in livestock such as cattle, sheep, and poultry. Coccidiosis can be extremely costly to the poultry industry, where overcrowded, stressful conditions promote the spread of disease and make treatment very difficult. Developmental problems are found in the young, and 20% of those infected die. In industrialised countries, many feeds contain a coccidiostat to prevent the disease.

Several drugs are available both for treatment and prevention of coccidiosis. A gene for a particular protein produced by the oocysts of the protozoan has been cloned so that purified antigen can be prepared for an effective vaccine. In vivo trials have been ongoing to determine the appropriate mixture of purified antigens for a vaccine that will boost immunity to coccidiosis.

Trypanosomiasis: More than 60 million cattle, as well as humans, pigs, domestic buffaloes, camels, horses, and small ruminants, in sub-Saharan Africa are susceptible to trypanosomiasis, a disease transmitted by the tsetse fly. Trypanosomiasis is difficult to treat since the single celled trypanosomes continually change their surface antigens during infection and can escape detection by the host's immune system. The most common method of preventing this disease is the routine use of chemical sprays or dips.

Research is focusing on ways to prevent infection by preparing vaccines from trypanosome components involved in the pathological process, such as the proteolytic enzymes used to degrade host molecules or proteins responsible for the suppression of the cattle immune system. Researchers are trying to isolate trypanosome-resistance genes in West African cattle so that transgenic trypanotolerant cattle can be produced. Host genes that inhibit parasite cell division may one day be identified and, through their increased expression, used to prevent the spread of the parasite within the host.

Theileriosis: Theileriosis, or East Coast fever, a deadly disease of cattle prevalent in 12 countries of east, central and southern Africa, is caused by several strains of tick-transmitted protozoans. Losses to African farmers exceed $170 million U.S. dollars. Once ingested by the tick, the parasite develops and finally forms sporozoites in the tick's salivary gland that are released into cattle bitten by infested ticks. Infected lymphocytes in the cattle become leukemic and eventually lympholysis occurs, usually causing death within one month of infection.

Ticks feeding on the blood of cattle infected with the protozoan transmit the disease to other cattle. At least 24 million cattle are at risk of infection, and high yielding exotic breeds are especially susceptible. Spraying or dipping cattle in a chemical insecticide has traditionally been used to control the disease. However, even if insect resistance can develop, the milk and meat may become contaminated with the pesticide, and chemicals can harm the environment. A crude preparation of live, infected ticks with an antibiotic has been used to stimulate cattle immune systems, but immunity is not always obtained and the antibiotic may not inhibit development of the parasite.

Recombinant DNA technology offers a means of producing new drugs for treatment, synthesising monoclonal antibodies for typing strains in geographical areas, and producing sporozoite antigen for vaccination in large quantities. If specific proteins expressed in the sporozoite stage of the parasite can be cloned and expressed in bacteria and insect cultures, these recombinant proteins in vaccine form may offer immunity in cattle.

Feed Utilisation and Nutrition

The shortage of feed in most developing countries and the increasing cost of feed ingredients mean that there is a need to improve feed utilisation. Aids to animal nutrition, such as enzymes, probiotics, single-cell proteins and antibiotics in feed, are already widely used in intensive production systems worldwide to improve the nutrient availability of feeds and the productivity of livestock. Gene-based technologies are being increasingly used to improve animal nutrition, either through modifying the feeds to make them more digestible or through modifying the digestive and metabolic systems of the animals to enable them to make better use of the available feeds.

Feeds derived from GM plants (a quarter of which are now grown in developing countries), such as grain, silage and hay, have contributed to increases in growth rates and milk yield. Genetically modified crops with improved amino acid profiles can be used to decrease nitrogen excretion in pigs and poultry. Increasing the levels of amino acids in grain means that the essential amino acid requirements of pigs and poultry can be met by diets that are lower in protein. Metabolic modifiers have also been used to increase production efficiency (weight gain or milk yield per feed unit), improve carcass composition (meat-fat ratio), increase milk yield and decrease animal fat.

The use of recombinant bovine somatotropin (rBST) in dairy cows increases both milk yield and production efficiency and decreases animal fat. In the USA, the use of rBST typically increases milk yield by 10% to 15%. Although trials conducted in developing countries have reported a similar percentage increase, this increase is not significant because of the low milk yields and the high cost-benefit ratio. However, rBST is being used commercially in 19 countries where the economic returns make its use worthwhile. A porcine somatotropin has been developed that increases muscle growth and reduces body-fat deposition, resulting in pigs that are leaner and of greater market value.

Limitation of Technological Applications

The application of new molecular biotechnologies and new breeding strategies to the livestock breeds used in smallholder production systems in developing countries is constrained by a number of factors. Over the last few decades, the green revolution has brought comparative prosperity to farmers with land, but the majority of farmers, who are landless or marginal farmers and subsist only on livestock, have been neglected and remain poor. The major constraints on applying biotechnologies are:

— the absence of an accurate and complete database on livestock and animal owners so that programmes can be implemented

— the biodiversity present within species and breeds in agro-ecological systems

- the fact that models of biotechnological intervention differ distinctly between developed and developing economies
- the fact that many animal species and breeds are unique to the developing world; each has its own distinct developmental, production, disease resistance and nutrient utilisation characteristics
- the lack of trained scientists, technicians and fieldworkers to develop and apply the technologies, both in the government and in the private sectors
- the absence of an interface between industry, universities and institutions, which is necessary to translate technologies into products
- the inability to access technologies from the developed world at an affordable price in order to make a rightful, positive and sustainable contribution to livestock production and the economic welfare of farmers
- the high cost of technological inputs such as materials, biologicals and equipment
- the failure to address issues of biosafety and to conduct risk analyses of new biologicals, gene products, transgenics and modified food items, and, above all
- the negligible investment in animal biotechnology.

The critical issues affecting livestock productivity have recently been re-examined. Research that aims to enhance productivity and sustainability should focus on improving livestock feeds and nutrition, improving animal health, managing natural resources relating to the livestock sector, assesing the impact of technological interventions, and strengthening the capacity of the national agricultural research systems of developing countries. Furthermore, the potential production capacity and contribution of livestock to the economy are still not being achieved in developing countries because the transfer, adaptation and adoption of technology is hampered by the lack of a clear policy for livestock development that is conducive to the introduction of new proven technology and by the lack of information flow from and to decision makers.

In developing countries, there is a wealth gap between urban and rural areas, which persists and may even be widening; the rural-urban divide also tends to be reflected in education and health indicators. In addition, women in rural (and urban) areas who are predominantly involved in animal husbandry have higher illiteracy rates than men. A survey of 21 African countries recently highlighted the substantial disparities in primary schooling between urban and rural areas, in favour of urban dwellers. Special attention must be given to the knowledge and information needed to enable rural people to apply biotechnology. There is a need to identify alternative delivery systems (beyond the State) for animal healthcare and to propose new roles for the state and the private sector in service delivery.

Capacity-building Activities

There is a great need to ensure that capacity is not just created but also is retained and enhanced. Capacity-building activities must be carried out at all levels: the awareness of policy and decision makers must be raised, the necessary legal and regulatory frameworks must be initiated, the technical and regulatory capacities must be enhanced and institutions may need to be overhauled. More importantly, it is necessary to assess and deploy competent operators and institutional capacity continuously so that, as biotechnology advances, the procedures required for its safe use can be constantly evaluated, upgraded and applied. This is a daunting task, but it can be achieved through firm commitment and partnerships.

Funding to Implement Technology

Developing and commercialising improved technologies in most developing countries has been the responsibility of the public sector, and technology has been disseminated freely. This situation will have to continue if superior genetics, diagnostics and vaccines are to be delivered. However, research and almost all commercial development of biotechnology in the developed world are being driven largely by the private sector. The global trends in funding for research and development and production do not address the concerns, needs and opportunities of the developing world. Developing countries are finding it increasingly expensive to access and use new technologies.

There is limited private- and public-sector investment in animal health and production, particularly in relation to modern biotechnologies that are 'resource hungry'. Although several discoveries have been made in laboratories in the developing world, in most cases these have not been converted into useful technologies or products. The key potential users -resource-poor often illiterate farmers with a limited knowledge base - do not feel that applying these technologies is worth the effort, cost and risk involved. This is mainly because there is no agency or industry that can scale up and package the technology. Also, in the developed world, there is an economic incentive to market biotechnological services and products; this is lacking in the developing world because of the limited purchasing power of resource-poor stakeholders.

Research in biotechnology in recent years has also been motivated by economic considerations, and little research is conducted in the developing world because of the probable lack of returns on the investment. For understandable reasons, current funding policies in developing countries focus on areas that will yield practical benefit in the short term. In determining future policy, policy-makers and funding bodies must not lose sight of the substantial benefits that can be gained in the longer term by investing in strategic research into vaccine development. Adequate multi-institutional (national and international) support through an international donor consortium is needed to develop cost-effective, cheap and easily adaptable biotechnological products. The amount spent by international agencies on animal biotechnology in developing countries is currently very low and constitutes only a small percentage of the total spending on agriculture.

6

Biotechnology for Food Security

Modern biotechnology is the application of (i) in vitro nucleic acid techniques, including recombinant deoxyribonucleic acid (DNA) and direct injection of nucleic acid into cells or organelles, or (ii) fusion of cells beyond the taxonomic family, that overcome natural physiological reproductive or recombination barriers, and that are not techniques used in traditional breeding and selection. The application of modern biotechnology to food production presents new opportunities and challenges for human health and development.

Recombinant gene technology, the most well-known modern biotechnology, enables plants, animals and microorganisms to be genetically modified (GM) with novel traits beyond what is possible through traditional breeding and selection technologies. It is recognised that techniques such as cloning, tissue culture and marker-assisted breeding are often regarded as modern biotechnologies, in addition to genetic modification. The inclusion of novel traits potentially offers increased agricultural productivity, or improved quality and nutritional and processing characteristics, which can contribute directly to enhancing human health and development.

From a health perspective, there may also be indirect benefits, such as reduction in agricultural chemical usage, and enhanced farm income, crop sustainability and food security, particularly in developing countries. The novel traits in genetically modified organisms (GMOs) may also, however, carry potential direct risks to human health and development. Many, but not all, genes and traits used in agricultural GMOs are novel and have no history of safe food use. Several countries have instituted guidelines or legislation for mandatory premarket risk assessment of GM food. At the international level, agreements and standards are available to address these concerns.

GMOs may also affect human health indirectly through detrimental impacts on the environment, or through unfavourable impacts on economic (including trade), social and ethical factors. These impacts need to be assessed in relation to the benefits and risks that may also arise from foods that have not been genetically modified. For example, new, conventionally bred varieties of a crop plant may also have impacts—both positive and negative—on human health and the environment.

Foods produced through modern biotechnology can be categorised as follows:

1. Foods consisting of or containing living/viable organisms, e.g. maize.
2. Foods derived from or containing ingredients derived from GMOs, e.g. flour, food protein products, or oil from GM soybeans.
3. Foods containing single ingredients or additives produced by GM microorganisms (GMMs), e.g. colours, vitamins and essential amino acids.
4. Foods containing ingredients processed by enzymes produced through GMMs, e.g. high-fructose corn

syrup produced from starch, using the enzyme glucose isomerase (product of a GMM).

Crop Breeding

Conventional breeding, especially of crops, livestock and fish, focuses principally on increased productivity, increased resistance to diseases and pests, and enhanced quality with respect to nutrition and food processing. Advances in cellular genetics and cell biology methods in the 1960s contributed to the so-called 'green revolution' that significantly increased varieties of staple food crops containing traits for higher yield and resistance to diseases and pests in a number of both developed and developing countries. A key driver of the green revolution was to improve the potential to provide sufficient food for all.

The intensification and expansion of agriculture brought about by these methods and agricultural systems have, however, also resulted in new forms of health and environmental risks through, for example, increased use of agrochemicals and intensified cultivation resulting in soil erosion. The development of molecular biology in the 1970s and 1980s introduced more direct methods for the analysis of genetic sequences and allowed the identification of genetic markers for desired traits. Such marker-assisted breeding methods are the basis of some current conventional breeding strategies.

Whereas modern methods of breeding have significantly increased crop yields over the past 50 years, the future potential of these methods is constrained by the limitations in the natural diversity of trait genotype within crop species and sexual-compatibility boundaries between crop types. To overcome these problems, a number of interested groups (scientists, farmers, governments, agricultural companies) have since the 1980s considered other means to achieve the objectives of improved yields, sustainable agricultural systems, and improvements in human and animal health and the environment. This includes the use of more modern methods to introduce novel traits, such as tolerance to drought, salt, or pests.

To achieve these objectives, various public and, more recently, private research programmes have aimed to improve the understanding of and links between crop performance and molecular genetics. With the development and use of recombinant DNA in the 1980s, a tool to overcome the limitation of species incompatibility was found. Modern biotechnology employs molecular techniques to identify, select and modify DNA sequences for a specific genetic trait (e.g. insect resistance) from a donor organism (microorganism, plant or animal), and transfer the sequence to the recipient organism so that it expresses this trait. Various transformation methods are used to transfer recombinant DNA into recipient species to produce a GMO.

For plants, these include transformation mediated by Agrobacterium tumefaciens (a common soil bacterium that contains genetic elements for infection of plants) and biolistics-shooting recombinant DNA placed on microparticles into recipient cells. The methods used in the transformation of various animal species include microinjection, electroporation and germ-line cells. The success rate of transformations in animals tends to be lower than in plants, and to vary from species to species, thus requiring the use of many animals.

Genetic modification is often faster than conventional breeding techniques, as stable expression of a trait is achieved using far fewer breeding generations. It also allows a more precise alteration of an organism than conventional methods of breeding, as it enables the selection and transfer of a specific gene of interest. However, with the present technology, in many cases it leads to random insertion in the host genome, and consequently may have unintended developmental or physiological effects.

However, such effects can also occur in conventional breeding and the selection process used in modern biotechnology aims to eliminate such unintended effects to establish a stable and beneficial trait. It should be noted that conventional breeding programmes directed by the molecular analysis of genetic markers are also of critical importance to modern plant and animal breeding. However, human and environmental health consequences of these techniques are not considered here.

GM Crops

At present, only a few GM crops are permitted for food use and traded on the international food and feed markets. These include herbicide- and insect-resistant maize (Btmaize), herbicide-resistant soybean, rape (canola) oilseed, and insect- and herbicide-resistant cotton (primarily a fibre crop, though refined cottonseed oil is used as food). In addition, several government authorities have approved varieties of papaya, potato, rice, squash, sugar beet and tomato for food use and environmental release. The latter crops, however, are currently grown and traded only in a limited number of countries, mainly for domestic consumption. The regulatory status of GM crops varies among the countries that permit their use and updates can be found on various web-sites, including those of the Organisation for Economic Co-operation and Development (OECD) and the International Centre for Genetic Engineering and Biotechnology (ICGEB).

In 2004, the estimated global area of commercially grown transgenic or GM crops was 81 million hectares, grown by 7 million farmers in 18 developed and developing countries. Seven countries grew 99% of the global transgenic crop area in 2004. During the nine-year period 1996 to 2004, herbicide tolerance has been the dominant trait introduced to commercial GM crops, with insect resistance second. In 2004, herbicide tolerance present in soybean, maize and cotton accounted for 72%, or 58.5 million hectares, of global GM plantings. Insect-resistant Bt crops accounted for 15.7 million hectares (20%), and 'stacked' genes (GM cotton or maize crops with both herbicide tolerance and insect resistance) accounted for 8% or 6.8 million hectares of the global transgenic area.

Virus-resistant crops, such as papaya (resistant to ringspot virus), potato (tolerant to potato virus Y and potato leaf roll virus) and yellow crookneck squash (resistant to watermelon mosaic virus) are commercially grown on a very small area in comparison. The two dominant GM crop/trait combinations in 2004 were: herbicide-tolerant soybean, 48.4 million hectares or 60% of the global total; and Bt maize, 11.2 million hectares, equivalent to 14% of the global area planted to transgenic crops.

The commercial introduction of transgenic crop plants with agronomic traits is often referred to as the first generation of transgenic plants. Further development of GM crops with agronomic traits is continuing, and production of a range of GM crops with enhanced nutritional profiles is also under way. Various novel traits are currently being tested in laboratories and field tests in a number of countries. Many of these second-generation GM crops are still in the development stage and are unlikely to enter the market for several years.

Fish and Livestock

In terms of food production, the application of modern biotechnology to livestock falls into two main areas: animal production and human nutrition.

Fish

The projected increasing demand for fish suggests that GM fish may become important in both developed and developing countries. Enhanced-growth Atlantic salmon containing a growth hormone gene from Chinook salmon is likely to be the first GM animal on the food market. These fish grow 3-5 times faster than their non-transgenic counterparts, to reduce production time and increase food availability. At least eight other farmed fish species have been genetically modified for growth enhancement. Other fish in which genes for growth hormones have been experimentally introduced include grass carp, rainbow trout, tilapia and catfish.

In all cases, the growth-hormone genes are of fish origin. To address some of the practical problems of aquaculture, research attempts are seeking to improve disease resistance by producing Atlantic salmon with a rainbow trout lysozyme cDNA. Lysozyme has antimicrobial properties against fish pathogens such as Vibrio, Aeromonas and Yersinia. Another type of antimicrobial protein (silk moth cecropin) is under investigation in catfish. This would improve catfish resistance to diseases such as enteric septicaemia.

The farming of carnivorous fish species, such as trout and salmon, has led to overfishing of sand eels and capelin. To tackle this problem, research is looking into the possibility of altering the metabolism of these species by improving their digestion of carbohydrates, to enable a shift to a more plant-based diet. Lack of cold tolerance in warm-water species such as the common carp and tilapia can lead to significant stock losses in winter. The suggestion of work in this area is to alter the molecular conformation of lipids, thus increasing membrane fluidity.

To extend the geographical range of fish farming, an antifreeze gene from one fish species is transferred to the species of interest. Although freeze-resistant strains of Atlantic salmon have been produced, the level of antifreeze protein secreted by the salmon was insufficient to have a significant impact on the freezing point of blood. The issues concerned in the identification of hazards and the assessment of risks that could be associated with the release of GM fish are still being addressed. One of these aspects involves the production of sterile GM fish to minimise the environmental risk of releasing them into wild populations.

Livestock and Poultry

Foods derived from GM livestock and poultry are far from commercial use. Several growthenhancing novel genes have been introduced into pigs that have also affected the quality of the meat, i.e. the meat is more lean and tender. This research was initiated over a decade ago, but owing to some morphological and physiological effects developed by the pigs, these have not been commercialised. Many modifications to milk have been proposed that either add new proteins to milk or manipulate endogenous proteins.

Recently, researchers from New Zealand developed GM cows that produce milk with increased levels of casein protein. Use of such protein-rich milk would increase the efficiency of cheese production. Other work aims to reduce the lactose content of milk, with the intent of making milk available to the population of milk-intolerant individuals. Other applications of genetic modification in animal production in the early stages of R&D include improvement of disease resistance, increased birth rates in sheep, altered sex ratio in poultry, increased egg production in poultry by creating two active ovaries, and improved feed conversion in the 'enviropig' (environmentally friendly pigs that excrete less phosphorus). Most of this work is still theoretical and therefore estimates of time frames for possible commercial introductions of any of these applications are unavailable.

Microorganisms

Currently, there are no known commercial products containing live genetically modified microorganisms (GMMs) on the market. In the United Kingdom, GM yeast for beer production has been approved since 1993, but the product was never intended to be commercialised. Other microorganisms used in foods (which are in the R&D phase) include starter fermentation cultures for various foods (bakery and brewing), and lactic acid bacteria in cheese.

R&D is also aimed at minimising infections by pathogenic microorganisms and improving nutritional value and flavour. Attempts have been made to genetically modify ruminant microorganisms for protecting livestock from poisonous feed components. Microorganisms improved by modern biotechnology are also under development in the field of probiotics, which are live microorganisms that, when consumed in adequate amounts as part of food, confer a health benefit on the host.

Many enzymes used as processing aids in food and feed production are derived through the use of GMMs. This means that the GM microorganisms are inactive, degraded or removed from the final product. GM yeasts, fungi and bacteria have been in commercial use for this purpose for over a decade. Examples include: alpha-amylase for bread-making, glucose isomerase for fructose production, and chymosin for cheese-making. Most of the microorganisms modified for food processing are derivatives of microorganisms used in conventional food biotechnology. GMMs are also permitted in a number of countries for the production of micronutrients, such as vitamins and amino acids used for food or dietary supplement purposes.

An example is the production of carotenoids (used as food additives, colourants or dietary supplements) in GM bacterial systems. In the future, complete metabolic pathways could be integrated in GM microorganisms, enabling them to produce new compounds. For animal husbandry, veterinary products such as bovine somatropin, used for increasing milk production, have been developed using genetic engineering. Bovine somatropin has been on the market in several countries for over a decade.

The technique of protein engineering aims at altering the genetic, and thus amino acid, sequence of enzymes. Hitherto, protein engineering has not been used extensively in enzyme production. R&D in this area aims to change enzyme characteristics, e.g. improve temperature or pH stability. Enzymatic processing often replaces existing chemical reactions. In many instances, this results in lowered energy consumption and less chemical waste.

Risk of GMOs and GM Foods

Introduction of a transgene into a recipient organism is not a precisely controlled process, and can result in a variety of outcomes with regard to integration, expression and stability of the transgene in the host.

When new foods (crop varieties, animal breeds or microorganisms) are developed by traditional breeding methods, they are usually not subject to specific pre- or postmarket risk or safety assessment by national authorities or through international standards. This is in contrast to requirements introduced for GMOs and GM foods. The concept of risk assessment of GMOs was first discussed at the Asilomar Conference in 1975. The discovery of recombinant DNA had raised concerns among researchers regarding the potential creation of recombinant viruses whose escape would threaten public health.

Fourteen months after a voluntary moratorium on research involving recombinant DNA techniques, guidelines for the physical and biological containment of riskier experiments were drafted and agreed. These guiding principles were the basis of the USA guidelines for research in modern biotechnology developed in 1976 by the National Institutes of Health Recombinant DNA Advisory Committee. Early regulatory requirements were intended to prevent the accidental release of microorganisms from research facilities.

In continuation of this, regulation for contained use and deliberate release of GMOs was developed, e.g. EU regulations in 1990. These guidelines elaborated a premarket humanhealth and environmental-safety assessment requirement for all GMOs and GM foods on the basis that they are novel and have no history of safe food or environmental use. Many countries have since established specific premarket regulatory systems requiring the rigorous assessment of GMOs and GM foods before their release into the environment and/or use in the food supply.

While many national regulatory bodies base their safety assessment of GMOs and GM foods on shared concepts, differences in regulatory systems have led to disagreements and confusion in their deployment. While the terms 'safety assessment' and 'risk assessment' are often used interchangeably in some literature, these are two clearly different, but interlinked processes. To provide international consistency in risk analysis of GMOs and GM foods which incorporates risk assessment, management and communication components, a number of international regulatory and standard-setting bodies have introduced uniform standards.

These include standards for human-health and environmental-safety assessment of GMOs and GM foods, and notification of their movement across national borders. The objective of uniform global standards for risk assessment would be challenging as countries are bound to reach different decisions on the scope of the assessment, particularly the resolution of whether or not to include social or economic aspects. International regulatory systems covering GM food safety (Codex Principles) and environmental safety (Cartagena Protocol on Biosafety) came into force in 2003.

The concept that allows for the comparison of a final product with one having an acceptable standard of safety is an important element of a GM food safety assessment. This principle was elaborated by FAO, WHO and OECD in the early 1990s and referred to as 'substantial equivalence'. The principle suggests that GM foods can be considered as safe as conventional foods when key toxicological and nutritional components of the GM food are comparable to the conventional food (within naturally occurring variability), and when the genetic modification itself is considered safe. However, the concept has been criticised by some researchers.

At a Joint FAO/WHO consultation on foods derived from biotechnology held in 2000, it was acknowledged that the concept of substantial equivalence contributes to a robust safety assessment, but it was also clarified that the concept should represent the starting point used to structure the safety assessment of a GM food relative to its conventional counterpart. The consultation concluded that a consideration of compositional changes should not be the sole basis for determining safety, and that safety can only be determined when the results of all aspects under comparison are integrated.

Influence of GM foods on Human Health

The Codex Alimentarius Commission adopted the following texts in July 2003: Principles for the risk analysis of foods derived from modern biotechnology; Guideline for the conduct of food safety assessment of foods derived from recombinant-DNA plants; and Guideline for the conduct of food safety assessment of foods produced using recombinant-DNA microorganisms. The last two texts are based on the Principles and describe methodologies for conducting safety assessments for foods derived from recombinant-DNA plants and microorganisms, respectively.

The premise of the Principles dictates a premarket assessment, performed on a case-by-case basis and including an evaluation of both direct effects (from the inserted gene) and unintended effects (that may arise as a consequence of insertion of the new gene). The Codex safety assessment principles for GM foods require investigation of:

(a) direct health effects (toxicity);

(b) tendency to provoke allergic reactions (allergenicity);

(c) specific components thought to have nutritional or toxic properties;

(d) stability of the inserted gene;

(e) nutritional effects associated with the specific genetic modification; and

(f) any unintended effects which could result from the gene insertion.

Codex principles do not have a binding effect on national legislation, but are referred to specifically in the Agreement on the Application of Sanitary and Phytosanitary Measures of the World Trade Organisation, and are often used as a reference in the case of trade disputes. The 2003 Expert consultation on the safety assessment of foods derived from GM animals, including fish formed the opinion that to further develop the risk-assessment process with current scientific knowledge, integrated toxicological and nutritional evaluations should be conducted in order to identify food-safety issues that may need further investigation.

Both evaluations combine data from the hazard identification and characterisation, and food intake assessment steps. It should be noted that such newly suggested further developments of the risk-assessment process have not yet been considered by Codex, and that the international principles and guidelines for risk analysis and safety assessment of foods derived from biotechnology are as accepted by Codex in 2003.

The potential direct health effects of GM foods are generally comparable to the known risks associated with conventional foods, and include, for example, the potential for allergenicity and toxicity of components present, and the nutritional quality and microbiological safety of the food. Many of these issues have not traditionally been specifically assessed for conventional food; but in one area—toxicity of food components—there is ample experience related to the use of animal experiments to test potential toxicity of targeted chemical components.

However, the intrinsic difficulty in testing whole foods, as opposed to specific components, in animal feeding experiments have resulted in the development of alternative approaches for the safety assessment of GM foods. The safety assessment of GM food follows a stepwise process aided by a series of structured questions. Factors taken into account in the safety assessment include:

— identity of gene of interest, including sequence analysis of flanking regions and copy number;

— source of gene of interest;

— composition of GMO;

— protein expression product of the novel DNA;

— potential toxicity;

— potential allergenicity; and

— possible secondary effects from gene expression or the disruption of the host DNA or metabolic pathways, including composition of critical macronutrients, micronutrients, antinutrients, endogenous toxicants, allergens and physiologically active substances.

Unintended effects, such as elevated levels of antinutritional or toxic constituents in food, have on occasion been characterised in conventional breeding methods, e.g. glycoalkaloid levels in potatoes. Organisms derived from conventional breeding methods, including tissue cultures, may have a somewhat enhanced possibility for genetic (and epigenetic—environmentally induced changes that affect the expression of a gene without changing the DNA sequence) instabilities, such as the activity of mobile elements and gene-silencing effects. These effects could increase the probability of unintended pleiotropic effects (affecting more than one phenotypic trait), e.g. increased or decreased expression of constituents or possibly modifications in expressed proteins, as well as epistasis (the interaction of the inserted gene with other genes).

It has been argued that random insertion of genes in GMOs may cause genetic and phenotypic instabilities but, as yet, no clear scientific evidence for such effects is available. A better understanding of the impact of natural transposable elements on the eukaryotic genome may shed some light on the random insertion of sequences. Gene expression in conventional and GM crops is subject to environmental influences. Environmental conditions such as drought or heat can stimulate some genes; turning the expression up or down.

The assessment of potential synergistic effects is necessary in the risk assessment of organisms derived from gene stacking, i.e. breeding of GMOs containing genetic constructs with multiple traits. Internationally agreed procedures for the assessment of such organisms are desirable. Unintended effects can be classified as insertional effects, i.e. related to the position of insertion of the gene of interest, or as secondary effects, associated with the interaction between the expressed products of the introduced gene and endogenous proteins and metabolites. There is common agreement that targeted approaches, i.e. the measuring of single compounds, is very useful and adequate to detect such effects, as has been done with conventionally bred products.

To enhance and improve the identification and analyses of these unintended effects, profiling methods have been suggested. This untargeted approach allows detection of unintended effects at the mRNA (microarray), protein (proteomic) and metabolite (metabolomic) level. It still remains to be seen which of these techniques (once validated) would be useful for routine risk-assessment purposes. Unintended effects were specifically addressed by the FAO/WHO Expert consultation on the safety aspects of genetically modified foods of plant origin and the Codex Principles for the risk analysis of foods derived from modern biotechnology. These consultations noted that there is a need to establish the consequences of natural baseline variations, the effects of growing conditions and environmental influences, and the ways to interpret safety-relevant data from profiling techniques.

Adequate methods for the assessment of potential, unintended effects need to be evaluated for specific GMOs case by case, where the assessment already aims to consider unintended toxic and antinutritional factors through analysis of proximal constituents and GM characteristics. As profiling methods are not in use in routine risk assessment, the second step in the comparative safety assessment has been suggested as a measure for identifying and characterising any unintended effects that may be associated with complex foods. Natural genetic transformation has been found to occur in different environments, e.g. in food. In addition, it has been shown that ingested DNA from food is not completely degraded by digestion, and that small fragments of DNA from GM foods can be found in different parts of the gastrointestinal tract.

As the consequences of horizontal gene transfer (HGT) may be significant in some human-health conditions, the potential for HGT needs to be part of the risk assessment of GM food. FAO/WHO consultations have also discussed the potential risks of gene transfer from GM foods to mammalian cells or gut bacteria. These panels have suggested that it may be prudent in a food-safety assessment to assume that DNA fragments survive in the human gastrointestinal tract and can be absorbed by either the gut microflora or somatic cells lining the intestinal tract.

It was agreed that the assessment needs to take into account a number of factors including, but not limited to, the specific characteristics encoded by the DNA sequences, the characteristics of the receiving organism, and the selective conditions of the local environment of the receiving organisms. Some scientists have pointed to the present methodological limitations of a comprehensive scientific evaluation of this problem (mainly because of estimations that only approximately 1% of naturally existing bacteria can be cultured, and therefore analysed). Discussion also addresses the consequences of a rare probability of a transfer event against the high numbers of bacteria and genes available for transfer.

The DNA construct used to change the genetic composition of a recipient organism should be considered within an assessment, especially if the gene or its promoter (e.g. cytomegalovirus promoter) has been derived from a viral source. Sequences unrelated to the target gene could be introduced as part of the construct. Inadvertent introduction of such sequences into the germ-line of a GM animal not only has the potential for creating unintended genetic damage, but can also contribute by recombination to the generation of novel infectious viruses.

A well-known example is the generation of a replication-competent murine leukaemia virus during the development of a vector containing a globin gene. The horizontal transfer of recombinant genetic material to microorganisms has demonstrated an enhanced stability of DNA under certain conditions. Natural transformation of DNA to bacteria involves the active uptake of extracellular DNA by bacteria in a status of competence or in rare, illegitimate recombination events. The probability of such an event occurring appears to be extremely low, and very much related to the genes, constructs and organisms in question.

Allergies and immune responses

Food allergies or hypersensitivities are adverse reactions to foods triggered by the immune system. Within the different types of reactions involved, non-immunological intolerances to food and reactions involving components of the immune system need to be differentiated. The former may invoke reactions such as bloating or other unpleasant reactions, but are thought not to involve the immune system and called 'food intolerances'. Allergic reactions to traditional foods are well known.

The major food allergens are proteins in and derived from eggs, fish, milk, peanuts, shellfish, including crustaceans and molluscs (e.g. clams, mussels and oysters), soy, tree nuts (e.g. almonds, Brazil nuts, cashews, hazelnuts/filberts, macadamia nuts, pecans, pine nuts, pistachios and walnuts) and wheat. Whereas the groups of main allergens are well known and advanced testing methods have been elaborated, traditionally developed foods are not generally tested for allergens before market introduction.

The application of modern biotechnology to crops has the potential to make food less safe if the newly added protein proves to cause an allergic reaction once in the food supply. A well-known case is the transfer of a gene encoding a known allergen, the 2S-Albumin gene from the Brazil nut, to a previously safe soybean variety. When the allergenic properties of the transgenic soybean were tested, sera from patients allergic to Brazil nuts cross-reacted with the transgenic soybean. For this reason, a commercial product was never pursued.

On the other hand, the introduction of an entirely new protein that has not been previously found in the food chain represents a different case. In the first case, guidelines for assessing foods with known allergens are clear. The second case is more difficult to assess because there is no definitive test to determine the potential allergenicity of a novel protein. Instead, several risk factors provide a rough guide as to the likelihood of allergenicity. Risk-assessment protocols for food allergy examine four elements:

1. allergenicity assessment (is the food or elements in the food a potential allergen);
2. dose response assessment (is there a safe concentration of the allergen);
3. exposure assessment (how likely is it that people will encounter the allergen); and
4. susceptible subpopulations (how do those prone to allergy react to this new food).

Elements of an allergenicity assessment include a comparison of the sequence of the transferred gene (including the flanking regions at the site of insertion) with sequence motifs of allergenic proteins from databanks, an evaluation of the stability of the newly expressed proteins against digestion, and animal and immune tests, as appropriate. Absence of sequence similarity with allergenic protein epitopes, and low stability under acidic or proteolytic conditions, do not preclude the presence of a potential allergen. There are proven incidents which have contradicted the general rules, e.g. where small modifications in a protein sequence determine allergenicity.

Allergenicity prediction using protein-sequence motifs identified from a new allergen database has been proposed as a new and superior strategy for identifying potential allergens. Some experts consider that the use of sera from polysensitised patients is important for the testing of allergenicity. Areas of improvement of risk assessment of allergens include mechanistic studies of animal models and genomic techniques. FAO/WHO expert panels have established protocols for evaluating the allergenicity of GM foods on the basis of the weight of evidence. The strategy adopted is applicable to foods containing a gene derived from either a source known to be allergenic or a source not known to be allergenic.

The panels have, however, discouraged the transfer of genes from known allergenic foods unless it can be demonstrated that the protein product of the transferred gene is not allergenic. These principles have been applied by many regulatory agencies assessing the safety of GM foods and have provided the basis for Codex guidelines for the safety assessment of foods derived from biotechnology. The cellular basis of immune responses is not completely understood, and a better understanding of the interaction of the immune system and foods in general is required in order to decipher whether specific GM foods may have impacts on the immune system apart from allergenicity. The impact of cell-mediated reactions (without involvement of immunoglobulin E antibodies) on hypersensitivity reactions elicited by foods is a matter of current research.

Genetically modified animals have mainly been produced for biomedical research purposes. To date, no GM food animals have been introduced onto international markets. But GM food animals such as fish can be expected in the near future. In principle, the assessment of food and feed safety for GM animals follows the general principles of the assessment of GMOs outlined above. However, the specificities of the introduction of transgenes into animals, often using viral constructs for introduction into the germ-line, need distinct consideration.

A 2003 report of the Pew Initiative on Food and Biotechnology reviewed techniques for the production, uses and welfare of GM animals, as well as safety aspects. The risk assessment of foods derived from GM animals needs to be undertaken, as for other GM foods, on a case-by-case basis. This includes an assessment of potential recombination of viral vectors used for transformation with wild-type viruses, especially in poultry, where potential incomplete digestion could lead to intestinal uptake of orally administered proteins, and an assessment of peptide expression that may have hormonal activity (e.g. in fish). The FAO/WHO expert consultation on the Safety assessment of foods derived from GM animals, including fish held in 2003 addressed the key issues for food safety and evaluated the extent of scientific knowledge with regard to hazard identification and characterisation unique to transgenic animals.

Safety aspects of foods produced with GMMs

The production of food additives or processing aids using GMMs, where the microorganism is not a part of the food, has become an important and generally well-accepted technology, with a significant number of such products on the market. Experience with the purification of proteins in the biomedical field suggests that well-standardised purification protocols are of central importance for the safety of these products.

Where the GMMs are a part of the food matrix (e.g. starter culture containing live or sterilised microbes), certain criteria were established in 2001 by a Joint FAO/WHO expert consultation on foods derived from biotechnology for assessing the risks that may be associated with the preparation of such foods. These include the genetic constructs (vectors) used in the GM microorganisms, the pathogenic potential of GMM, and the detrimental effects of a potential gene transfer (considering a higher incidence for gene transfer and the various mechanisms involved).

For GMMs used in foods (e.g. in fermented foods or in functional food preparations), the ensuing risk assessment ought to focus on the effects of a possible interaction between the GMM and endogenous intestinal microflora and the potential immune-stimulatory or immunomodulatory effects of the microorganisms in the event the gastrointestinal tract is colonised. Small regulatory elements derived from viral DNA are commonly used to drive the expression of transgenes in GMOs. Viral-DNA constructs are sometimes used as transgenes to establish resistance against viral pests, as they express viral proteins that confer viral resistance on plants.

Some scientists suggest that the potential interaction of GM viral constructs with related wild-type viruses needs to be part of the risk assessment, to evaluate the potential of new viral pest strains evolving through mechanisms of recombination. Insertion of viral vectors into functionally important genes of recipient patients in the field of biomedicine has been reported, and whereas such vectors are not commonly used in food production, this evidence indicates the limited understanding of mechanisms directing insertion of genetic constructs.

Biopharming

The potential to produce human proteins in animals has resulted in great interest in new possibilities for human health, but also in efforts to establish appropriate risk-assessment methods. The biosafety aspects of molecular 'farming' (or 'pharming') can be divided into two major groups: the potential spread of transgenes; and the potential negative effects of the expressed protein on the environment and the consumer. Practices and guidelines ensuring effective separation of 'biopharming' are being investigated. Experts agree that the risk assessment should ensure that proteins designed to produce pharmaceutical products, e.g. in the animal's milk, cannot find their way to other parts of the animal's body, possibly causing adverse effects.

Environmental Safety and GMOs

In many national regulations, the elements of environmental risk assessment (ERA) for GM organisms include the biological and molecular characterisations of the genetic insert, the nature and environmental context of the recipient organism, the significance of new traits of the GMO for the environment, and information on the geographical and ecological characteristics of the environment in which the introduction will take place. The risk assessment focuses especially on potential consequences for the stability and diversity of ecosystems, including putative invasiveness, vertical or horizontal gene flow, other ecological impacts, effects on biodiversity and the impact of the presence of GM material in other products.

Different approaches in the ERA regulations of different countries have often resulted in different conclusions on the environmental safety of certain GMOs, especially where the ERA focuses not only on the direct effects of GMOs, but also addresses indirect or long-term effects on ecosystems, e.g. impact of agricultural practices on ecosystems. Internationally, the concept of 'familiarity' was developed also in the concept of environmental safety of transgenic plants. The concept facilitates risk/safety assessments, because to be familiar means having enough information to be able to make a judgement of safety or risk.

Familiarity can also be used to indicate appropriate management practices, including whether standard agricultural practices are adequate or whether other management practices are needed to manage the risk. Familiarity allows the risk assessor to draw on previous knowledge and experience with the introduction of plants and microorganisms into the environment and informs appropriate management practices. As familiarity depends also on knowledge of the environment and its interaction with introduced organisms, the risk/safety assessment in one country may not be applicable in another country.

Currently, the Cartagena Protocol on Biosafety (CPB) of the Convention on Biological Diversity is the only international regulatory instrument which deals specifically with the potential adverse effects of GMOs (known as living modified organisms (LMOs) under the Protocol) on the environment, taking also into account effects on human health. The Protocol covers transboundary movements of any GM foods that meet the definition of an LMO. Annex III of the Protocol specifies general principles and methodology for risk assessment of LMOs.

The Protocol establishes a harmonised set of international rules and procedures designed to ensure that countries are provided with the relevant information, through the information exchange system called the Biosafety Clearing House. This Internet-based information system enables countries to make informed decisions before agreeing to the importation of LMOs. It also ensures that LMO shipments are accompanied by appropriate identification documentation.

While the Protocol is the key basis for international regulation of LMOs, it does not deal specifically with GM foods, and its scope does not consider GM foods that do not meet the definition of an LMO. Furthermore, the scope of its consideration of human-health issues is limited, given that its primary focus is biodiversity, in line with the scope of the Convention itself. Consequently, the Protocol alone is not sufficient for the international regulation of GM foods.

In the future, GMOs may gain wider approval for environmental release, either with or without approval to enter them in the human food supply. In such situations, it will be important to consider whether or not to apply postmarket monitoring for unexpected environmental spread of the GMOs and their transgenes) that may pose food safety hazards. Methods for detection of such GMOs and their transgenes in the environment are likely to involve application of two well-established bodies of scientific methodologies: (1) diagnostic, DNA-based markers; and (2) sampling protocols that are adequate (in terms of statistical power) and cost-effective.

However, there is a need to fully develop appropriate protocols for application of these methods to postmarket detection of environmental spread of GMOs and their transgenes. Monitoring can also be helpful to assure confinement of GMOs during R&D. Postmarket monitoring (or surveillance) of GM foods with respect to direct human-health impacts has been raised at international conferences and within the Codex Alimentarius Commission. Opinions regarding such monitoring vary from neither necessary nor feasible, to being essential to support and improve the results of a risk assessment and enable an early detection of uncharacterised and unintended hazards.

Some have suggested that monitoring of potential long-term effects of GM foods with significantly altered nutritional composition should be mandatory. The Expert consultation on the safety assessment of foods derived from GM animals held in 2003 identified a need for postmarket surveillance, and therefore a product-tracing system, for:

— confirmation of the (nutritional) assessments made during the premarket phase;

— assessment of allergenicity or long-term effects; and

— unexpected effects.

The issue of postmarket surveillance is closely related to risk characterisation. In general, potential safety issues should be addressed adequately through premarket studies, as the potential of postmarket studies is currently very limited. Postmarket surveillance could be useful in certain instances where clear-cut questions require, for instance, a better estimate of dietary exposure and/or the nutritional consequences of GMO-derived food.

Tools to identity or trace GMOs or products derived from GMOs in the environment or food-chain are a prerequisite for any kind of monitoring. Detection techniques (such as polymerase chain reaction; PCR) are in place in a number of countries to monitor the presence of GMOs in food, to enable the enforcement of GM labelling requirements, and to monitor effects on the environment. Attempts to standardise analytical methods for tracing GMOs have been initiated.

Capacity Building

The United Nations and international development agencies coined the term 'capacity building' in the early 1990s after an appraisal of the development-assistance programmes in developing countries. Within that context, the term has come to mean different things to different people. In 1997, a World Bank progress report on Africa defined it as: "...an investment in people, institutions and practices that will, together, enable countries in the region to achieve their development objectives". Capacity building is a four-step process involving a needs assessment, strategic planning to change the situation, training of personnel to implement the changes, and an evaluation of the results.

A United Nations Development Programme (UNDP) report describes capacity building as a continuous process that ought to take place at various levels: individual, institutional and societal. The first two levels involve expanding local knowledge and skills. At the societal level, it is about creating opportunities to engage the trained individuals to their fullest potential. All three levels are interdependent and need to be pursued concurrently in order to achieve the maximum benefit. In a report published in 2000, UNDP acknowledged that due to the different levels of development among countries, some countries may never be in a position to deploy cutting-edge technologies. Nevertheless, these countries need local expertise to understand and adapt technologies for national use, consistent with their development goals.

Capacity Needs

Food safety is attracting increased attention because of its implications for public health. In general, food control systems in developing countries are poorly developed, and less organised than in most developed countries. Their overall capacity needs in terms of food safety can be summarised as follows:

— basic infrastructure;

— national food control strategy;

— food legislation and regulatory framework;

— food inspection services;

— food control laboratories and equipment; and

— implementation of food quality and safety assurance systems.

The work in food safety is multidimensional, and there are frequently several food laws under the authority of different agencies. In many countries, effective food control is undermined by the existence of fragmented legislation, multiple jurisdictions and weaknesses in surveillance, monitoring and enforcement mechanisms. Food-safety legislation developed specifically for the safety of GM foods should be integrated within the existing food laws, taking into account the special risk-management requirements. In order to make informed decisions on the safety of GMOs and GM foods, governments need substantial human and institutional resources in the disciplines required for assessing the risks for the environment and for human food presented by GMOs.

Developing countries have limited expertise in the required fields of science, as biotechnologists in these countries are generally engaged in research and therefore mostly unavailable to the regulatory bodies and as policy-makers. In most developing countries, those same scientists sit on national biosafety committees, and are involved in both risk assessment and policy-making. There are three vulnerabilities in this scenario:

(a) when developers are also risk assessors, the potential for conflict of interest is magnified;

(b) because most members of the national biosafety committee are recruited on a voluntary basis, they do not devote too much time to this responsibility; and

(c) because membership of the national biosafety committee generally rotates, there is no continuity in the capacity gained through experience.

While many developed countries have adopted mechanisms to govern modern biotechnology, most developing countries are either in the process of developing national biosafety frameworks or are yet to start the process. To date, no more than 10 developing countries have implemented national biosafety laws. A further 20-30 are in a state of transition whereby some or all elements are at different stages of development.

A few developing countries that permit the commercial planting of crops derived from modern biotechnology have modest capacities to implement a regulatory framework. Where national biosafety frameworks are in place, they vary between countries according to national priorities and statutory structures. In addition, the different social conditions that prevail in different countries make it difficult to determine the appropriate regulatory systems that should be enforced by developing countries. Notwithstanding the diversity, a number of elements are essential and form the core of many national frameworks:

— national policy and strategy;

— regulatory framework consisting of regulations and guidelines;

— mechanism for handling applications and issuing permits;

— system for enforcement; and

— system for information dissemination.

The impetus to establish regulatory frameworks for biosafety seems to be a significant factor in determining the process whereby they are developed. In some cases, scientists have raised interest in regulating local research, while in others the trigger may have come from multinational companies seeking to continue seed production in the Southern Hemisphere during the northern winter months. Recently, the importation of food aid has triggered some form of regulation in those countries that have been faced with food shortages. Many countries with regulatory systems have developed and implemented these systems in a stepwise fashion, usually in response to an immediate demand.

The first step has involved the establishment of voluntary guidelines to set in process a structured progression of the regulatory framework. The guidelines initially set the principles for safety in laboratory practices, which are later adapted to ensure environmental safety for enabling field trials. The advantage of guidelines is that revising and incorporating new information requirements in line with an evolving technology can be done very swiftly. However, guidelines are voluntary and compliance cannot be enforced unless supported by regulations.

FOOD SECURITY

The official definition of food security, adopted at the World Food Summit of 1996, states: "Food security exists when all people, at all times, have physical and economic access to sufficient, safe and nutritious food to meet their dietary needs and food preferences for an active and healthy life."

The causes of food insecurity involve a complex interplay of economic, social, political and technical issues. An analysis of this interplay should determine the potential solution and best approach for a given population group. The issue for some communities is being able to produce sufficient food. For others, lack of money to purchase a wider selection of foods is the problem. Food insecurity and poverty are strongly correlated. The Swedish International Cooperation Agency (SIDA) defines poverty as a three-fold deficiency: a lack of security, ability and opportunity. Poverty is the main cause of food insecurity, and hunger is also a significant cause of poverty.

Hunger is not only about quantity—it goes hand-in-hand with malnutrition. Food insecurity and malnutrition impair people's ability to develop skills and reduce their productivity. A lag in farm productivity is closely associated with rural poverty and hunger. Food insecurity nevertheless is a reality experienced by the vulnerable in all societies and in all countries, developed and developing. In developed countries, the problem of food security is often a reflection of affordability and accessibility through conventional channels.

Food security for the rural poor in developing countries is about producing or securing enough to feed one's household and being able to maintain that level of production year after year. Hunger and malnutrition increase susceptibility to disease and reduce people's ability to earn a livelihood. In instances where hunger is related to household income, improving food security by ensuring access to food or increasing the purchasing power of a family is essential. Providing poor communities with the skills to improve conditions in an economically and ecologically sustainable manner creates a window of opportunity to alleviate poverty at the subsistence-farming level and, on a larger scale, by having an impact on the economic development of the country.

Historically, increased food production in the developing countries can be attributed to the cultivation of more land rather than to the deployment of improved farming practices or to the application of new technologies. By its very nature, agriculture threatens other ecosystems, a situation that can be exacerbated by over-cultivation, overgrasing, deforestation and bad irrigation practices. However, increased demands for food in Asia, Europe and North Africa have to be met by increasing yields because most land in these areas is already used for agriculture.

The potential to expand agricultural land exists in Latin America and sub-Saharan Africa only, where much of the remaining land is marginal for agricultural expansion. The implication, therefore, is that the increase in food production needed to feed the world's growing population can only be met by increasing the amount of food produced per hectare. Recognising the extent of environmental degradation caused mainly by human activities, the multilateral agreements that arose from the UNCED meeting of 1992 were intended to address the compromised food-security situation on a global scale.

One such agreement is the United Nations Convention to Combat Desertification. This agreement promotes the implementation of practices intended to reverse desertification for sustainable land use and food security. As the more affluent developed countries tend to produce more food, some argue that redistribution of these surpluses could feed the escalating populations of developing countries. Redistribution, however, requires policy changes that may be impossible to implement on a global scale. Therefore, a substantial proportion of the food demands of developing countries will have to be met by the agricultural systems in these countries.

Enabling a consistent and sustainable supply of food will require an overhaul in the production processes and the supporting infrastructure. Finding solutions to declining crop yields requires an effort that will improve the assets on which agriculture relies; namely, soils, water and biodiversity. Transforming the agricultural systems of rural farmers by introducing technologies that integrate agro-ecological processes in food production, while minimising adverse effects to the environment, is key to sustainable agriculture. In addition, increases in crop yield must be met with the use of locally available low-cost technologies and minimum inputs without causing damage to the environment.

Global food productivity is undergoing a process of rapid transformation as a result of technological progress in the fields of communication, information, transport and modern biotechnology. A general observation is that technologies tend to be developed in response to market pressures, and not to the needs of the poor who have no purchasing power. As agriculture is the main economic activity of rural communities, optimising the levels of production will generate employment and income, and thus uplift the wealth and well-being of the community.

Improving agricultural production in developing countries is fundamental to reducing poverty and increasing food security. Investment to raise agricultural productivity can be achieved through the introduction of superior technologies such as better-quality seeds, crop rotation systems etc.. It is argued, however, that the adoption of earlier agricultural technologies has led to the emergence of more virulent strains of pests, pathogens and weeds, soil deterioration and a loss of biodiversity. The Green Revolution, in particular, focused on wheat and rice—not much attention was paid to staple crops such as sorghum, cassava or millet.

Also, the seeds and fertilisers required to grow the higher-yielding varieties were expensive and therefore not accessible to all. Reaffirming support for the principles agreed upon at the UNCED, the United Nations Millennium Development Goals have set a road map for protecting the environment. Embraced in these time-bound goals is a new development ethic that demands sustainability in a framework where progress is measured in terms of actions reconciling the economic and ecological factors of food production for the benefit of present and future generations?

A secure food system is one in which the ecological resources on which food production depends allow for their continued use, with minimum damage for present and future generations. In other words, food security and sustainable agriculture are interlinked and both are central to the concept of sustainable development. The FAO Anti-hunger Programme reported that increasing investment in agriculture and rural development can reduce hunger.

The production problems experienced by farmers vary between countries and communities, and technological solutions need to be relevant to those circumstances, i.e. one solution will not be suitable everywhere. The potential of some of these technologies has been demonstrated in various world regions—for instance, agro-ecological improvement programmes involving:

- better harvesting and conservation of water, even in rain-fed environments;
- a reduction of soil erosion by adopting zero tillage combined with the use of green manure and herbicides as in Argentina and Brazil; and
- pest and weed control without pesticide or herbicide use, e.g. Bangladesh and Kenya, has been well tested and established.

Indeed, such programmes are now widely accepted as being at the core of sustainable agriculture. The communities that participated in these projects were able to transform food production through the use of resource-management strategies that focused on improving the soil by growing leguminous crops and applying agro-forestry, zero tillage and green manure. These and other projects have proved that the sustainability of any farming practice and the conditions under which production can be maintained at reasonable levels cannot be predicted with absolute certainty.

Some regions may be better able to transfer high-yielding technologies with varying degrees of success. The uptake of new production systems has proved successful where the programmes have included the participation of entire communities and have not been introduced to isolated groups of farmers. Producing nutritionally enhanced properties in staple crops eaten by the poor could reduce the burden of disease in many developing countries.

Scientists at the International Crops Research Institute for the Semi-Arid Tropics (ICRISAT, India) have developed a pearl millet variety enhanced with betacarotene. The trait naturally occurs in two Burkina Faso millet lines from which it was transferred by conventional breeding methods. Genetic modification of japonica rice with a ferritin gene has not given superior results compared to rice with an 80% increase in iron density produced by conventional plant breeding at the International Rice Research Institute. Research and technology alone will not drive agricultural growth.

Inadequate infrastructure and poorly functioning markets tend to exacerbate the problem of food insecurity. The cost of marketing farm produce can be prohibitive for small-scale farmers, as their isolation prevents the link between agricultural and non-agricultural activities among adjacent villages and between rural and urban areas. Building roads in rural areas is vital to facilitating growth, trade and exchange of farm and non-farm products in rural communities, even those that can adequately feed themselves.

For instance, government investment in irrigation projects, storage and transport facilities, roads connecting villages to larger markets in the rural areas of China and India, has made an impressive impact on employment and productivity and ultimately provided opportunities for poverty alleviation in the affected areas. According to UNDP, basic thresholds in roads, power, ports and communications must be reached in order to sustain growth.

Role of Modern Biotechnology in Food-security

The Convention on Biological Diversity dictates the use and application of relevant technologies as a means of achieving the objectives of conservation and sustainable use with specific reference to biotechnology. Modern biotechnology is purported, from a technical perspective, to have a number of products for addressing certain food-security problems of developing countries. It offers the possibility of an agricultural system that is more reliant on biological processes rather than chemical applications.

The potential uses of modern biotechnology in agriculture include: increasing yields while reducing inputs of fertilisers, herbicides and insecticides; conferring drought or salt tolerance on crop plants; increasing shelf-life; reducing postharvest losses; increasing the nutrient content of produce; and delivering vaccines. The availability of such products could not only have an important role in reducing hunger and increasing food security, but also have the potential to address some of the health problems of the developing world. Achieving the improvements in crop yields expected in developing countries can help to alleviate poverty: directly by increasing the household incomes of small farmers who adopt these technologies; and indirectly, through spill-overs, as evidenced in the price slumps of herbicides and insecticides.

Indirect benefits as a whole tend to have an impact on both technology adopters and non-adopters, the rural and urban poor. Indeed, some developing countries have identified priority areas such as tolerances to alkaline earth metals, drought and soil salinity, disease resistance, crop yields and nutritionally enhanced crops. The adoption of technologies designed to prolong shelf-life could be valuable in helping to reduce postharvest losses in regionally important crops. Prime candidates in terms of crops of choice for development are the so-called 'orphan crops', such as cassava, sweet potato, millet, sorghum and yam.

Multinationals have found no incentive to develop these crops and have instead invested in marketable crops with high profit returns. This strategy is intended to target wealthier farmers in temperate-zone countries with the financial capacity and tradition of supporting new seed products. However, here is a potential for multinational companies to develop crops grown largely in developing countries. The investment costs are low and the potential markets considerably large.

While some public-sector research institutes in developing countries are forging ahead with the application of modern biotechnology, a small number are supported by government policy and therefore follow a defined agenda. Still other governments believe that the risks (safety, environmental and/or economic) associated with modern biotechnology outweigh the benefits. Currently, the many promises of modern biotechnology that could have an impact on food security have not been realised in most developing countries.

In fact, the uptake of modern biotechnology has been remarkably low owing to the number of factors that underpin food security issues. In part, this could be because the first generation of commercially available crops using modern biotechnology were modified with single genes to impart agronomic properties with traits for pest and weed control, and not complex characteristics that would modify the growth of crops in harsh conditions. Secondly, the technologies are developed by companies in industrialised countries with little or no direct investment in, and which derive little economic benefit from, developing countries. Thirdly, many developing countries do not have the necessary biosafety frameworks to regulate the products of modern biotechnology.

Although current commercial GM crops are not designed to address the specific issues of developing countries, their adoption has shown that they can be relevant in some developing countries—for example, the planting of herbicide-tolerant soybeans in Argentina and Bt cotton as a cash crop by resource-poor farmers in China and South Africa. There is little information on the economic costs associated with R&D of products of modern biotechnology, or on the impact of their introduction on production costs.

An in-depth analysis of the short- and long-term economic and social costs and benefits is necessary. Qaim and Zilberman report that farmers in Argentina that adopted herbicide-tolerant soybeans reduced per hectare production costs through the reduced number of herbicide applications, and thereby increased total factor productivity by 10%. On average, the Bt cotton farmers in China reduced pesticide spraying for the Asian bollworm by 70%, producing a kilogram of cotton at 28% less cost than the non-Bt farmers.

These benefits have had a significant impact on the agronomic, environmental, health and economic situations of approximately 5 million resource-poor farmers over eight provinces. Similarly, farmscale trials in China of GM rice containing genes which make them resistant to insect larvae that devastate rice crops showed 80% less pesticide use and yields increased by 6-9%. In addition, farmers who grew the GM varieties suffered less pesticide-induced illness than those growing the old varieties. Developing countries with limited financial and human resources need to find the right balance for investing in conventional and modern biotechnology research programmes.

While alliances with the private sector may contribute to the search for new technologies, the public sector needs to focus on crops and traits in which the former may be unwilling or unable to invest. The extent to which priority is given to modern biotechnology over other research methods should be linked to a country's agricultural priorities and objectives as well as to its environmental concerns. Ultimately, investment in interventions that support good governance, the development of rural infrastructure and market access is required before any of the promises of modern biotechnology can be realised. In general, policies that stimulate economic growth and target poverty reduction may have significant bearing on the health and well-being of the population.

7

Applications of Marine Biotechnology

Applications of biotechnology are found in health care, environmental remediation and the production of energy, food and fine chemicals. Bioprocess engineering has been the key to success in the commercialisation of biotechnology, especially with respect to biopharmaceuticals. In marine biotechnology, both new and existing biotechnological techniques are developed and applied to organisms from marine sources. For marine biotechnology, bioprocess engineering represents the link between discovery and commercialisation. The world's oceans represent one of the largest untapped biological resources.

Appropriately, the United Nations has designated 1998 as the 'Year of the Ocean' in recognition of the importance of oceans to the wellbeing of this planet and the need for proper use of their resources for sustainable development. Oceanic organisms are of enormous scientific interest, for two major reasons: First, they constitute a major share of the Earth's biological resources. Second, marine organisms often possess unique structures, metabolic pathways, reproductive systems, and sensory and defence mechanisms because they have adapted to extreme environments ranging from the cold polar seas at 2°C to the great pressures of the ocean floor, where hydrothermal fluids spew forth.

Most of the major classes of the Earth's organisms are primarily or exclusively marine, so the oceans represent a source of unique genetic information. The diversity of marine life points to a myriad of new bioproducts waiting to be discovered and developed commercially through engineering research and a systems approach that spans activities from the identification of new marine bioproducts through production, separation, formulation and delivery. Numerous potential pharmaceuticals (e.g. bryostatin, halichondrin), valuable biopolymers (e.g. chitin), compounds that can be used in the anti-fouling industry (as an alternative to toxic chemicals like organotins), products for the food industry (poly-unsaturated fatty acids) and products for the cosmetic industry (e.g. phycobiliproteins) have been found in and isolated from marine organisms.

In addition, marine microorganisms can be used to produce energy (biohydrogen). In order to stimulate research that leads to commercial exploitation of these natural resources, marine organisms need to be cultivated in vitro for the rational and sustainable exploitation of the sea as a source of food and natural products.

In recent years increasing funds have been devoted to marine biotechnology research; the United States, for example, spent some $55 million in 1995. Develop a fundamental understanding of the genetic, nutritional, and environmental factors that control the production of primary and secondary metabolites in marine organisms, as a basis for developing new and improved products. Identify bioactive compounds and determine their mechanisms of action and natural function, to provide models for new lines of selectively active materials for application in medicine and the chemical industry.

Develop bioremediation strategies for application in the world's coastal oceans, where multiple uses-including wastewater disposal, recreation, fishing, and aquaculture-demand prevention and remediation of pollution; and develop bioprocessing strategies for improving sustainable industrial processes. Use the tools of modern biotechnology to improve the health, reproduction, development, growth, and overall well-being of cultivated aquatic organisms; and promote the interdisciplinary development of environmentally sensitive, sustainable systems that will enable significant commercialisation of aquaculture. Improve understanding of microbial physiology, genetics, biochemistry, and ecology in order to provide model systems for research and production systems for commerce, and to contribute to understanding and conservation of the seas.

New and Improved Marine Products

Most major groups of living organisms are primarily or exclusively marine. Tropical marine environments harbor an especially wide diversity of animals and plants. Many marine organisms are sessile and must employ sophisticated methods to compete for a place to anchor. This characteristic is reflected in part by a metabolism that produces enormously diverse bioactive products-many with no terrestrial counterparts. Recent research has uncovered unicellular and multicellular microorganisms that are unique to the marine world.

Indeed, "... it seems clear that marine microorganisms are emerging as significant chemical resource". Before this vast biological resource can be developed and exploited, however, scientists must find answers to fundamental questions concerning as to what are actually the nutritional requirements of marine microorganisms and the amounts and characteristics of chemicals they produce.

Therapeutic Natural Products

We already have an idea about the various Therapeutic natural products found in terrestrial plants and microorganisms were the basis of early drug to be explored and used as sources of new drugs, albeit with declining rates of success. By contrast, recent trends suggest that exploration and use of marine resources, especially microorganisms (bacteria, fungi, and micro algae), for drug development has barely begun. Many bioactive substances from the marine environment already have been isolated and characterised, with great promise for the treatment of human diseases.

The compound manoalide from a Pacific sponge, for example, has spawned more than 300 chemical analogs, with a significant number of these going on to clinical trials as anti-inflammatory agents. To date, exploitation of natural agents from the sea has been hindered by problems with limited or sporadic distribution and production. Much more research must be conducted to determine what seasonal factors and life cycle or reproduction states are linked with natural production of an agent.

Factors influencing production may include diet, physical and chemical conditions, distribution by phylogenetic affiliation, geographic location, water depth, or associations with symbiotic microorganisms. Knowledge of these factors will be important in developing methods for producing selected metabolites, either from whole organisms or in vitro from cell or tissue cultures of plants and animals. Many of these compounds are very large and complex molecules, requiring very elaborate biochemical processes; as a result, it can be difficult to synthesise them or clone all the genes for standard production through fermentation.

Biomolecular Materials

Enzymes

Enzymes produced by marine bacteria are important in biotechnology due to their range of unusual properties. As far as the enzymes are concerned some are saltresistant, a characteristic that is often advantageous in industrial processes. The extra cellular proteases are of particular importance and can be used in detergents and industrial cleaning applications, such as in cleaning reverse-osmosis membranes. Vibrio species have been found to produce a variety of extracellular proteases.

Vibrio alginolyticus produces six proteases, including an unusual detergent-resistant, alkaline serine exoprotease. This marine bacterium also produces collagenase, an enzyme with a variety of industrial and commercial applications, including the dispersion of cells in tissue culture studies. Other research has demonstrated the presence in algae of unique haloperoxidases (enzymes catalysing the incorporation of halogen into metabolites). These enzymes could become valuable products, because halogenation is an important process in the chemical industry.

Japanese researchers have developed methods to induce a marine alga to produce large amounts of the enzyme superoxide dismutase, which is used in enormous quantities for a range of medical, cosmetic, and food applications. An unusual group of marine microorganisms from which enzymes have been isolated are the hyperthermophilic archaea (previously called archaebacteria), which can grow at temperatures over

100°C and therefore require enzyme systems that are stable at high temperatures.

Archaea typically are found in extreme environments, such as hot springs, animal guts, hydrothermal vents, sewage sludge digesters, and hypersaline habitats, including the Great Salt Lake. Thermostable enzymes offer distinct advantages, many still to be discovered, in research and industrial processes. Thermostable DNA-modifying enzymes, such as polymerases, ligases, and restriction endonucleases, already have important research and industrial applications. Hot springs in Yellowstone National Park provided the first archaeon (Thermus aquaticus) from which thermostable DNA polymerases were isolated. These novel enzymes (the Taq® polymerases) became the basis for the polymerase chain reaction (PCR), a useful technique for studying genetic material.

In 1989, thermostable DNA polymerase was designated Molecule of the Year by Science magazine. Comparable enzymes continue to be discovered. Most enzymes involved in the primary metabolic pathways of thermophilic bacteria and archaea are dramatically more thermostable than are their counterparts living at moderate temperatures. Expanded study of enzymes from thermophilic marine microorganisms will contribute to the understanding of mechanisms of enzyme thermostability and should enable the identification of enzymes suitable for industrial applications as well as modification of enzymes to enhance thermostability.

Marine Biopesticide

Natural marine products have the potential to replace chemical pesticides and other agents used to maximise crop yields and growth. Continued Federal support for R&D in this area is likely to result in useful natural pesticides that would provide greater specificity and fewer harmful side effects than do conventional synthetic agents. An example of a marine biopesticide in use today is Padan, which was developed from a bait worm's toxin known to ancient Japanese fishermen. This natural pesticide has demonstrated activity against larvae of the rice stem borer, the rice plant skipper, and the citrus leaf miner, among other pests.

More recently, scientists in Montana discovered novel compounds in marine algae and marine sponges containing symbiotic microorganisms. These compounds promoted growth and stimulated germination and increased root and coleoptile lengths in test plants. Several sponge and nudibranch species produce terpenes, a broad class of aromatic compounds used in solvents and perfumes and known to deter feeding by fish. Extracts derived from this same sponge and nudibranch species also demonstrated powerful insecticidal activity against two species, grasshoppers and the tobacco hornworm.

Biomonitors

Marine organisms can provide the basis for development of biosensors, bioindicators, and diagnostic devices for medicine, aquaculture, and environmental monitoring. One type of biosensor employs the enzymes responsible for bioluminescence. The lux genes, which encode these enzymes, have been cloned from marine bacteria such as Vibrio fischeri and transferred successfully to a variety of plants and other bacteria. The lux genes typically are inserted into a gene sequence, or operon, that is functional only when stimulated by a defined environmental feature.

The enzymes responsible for toluene degradation, for example, are synthesised only in the presence of toluene. When lux genes are inserted into a toluene operon, the engineered bacterium glows yellow-green in the presence of toluene. This genetically engineered system "reports" that biodegradation of a specific chemical, in this case toluene, is proceeding. Another type of biomonitor that holds great promise is the gene probe, which can be used to identify organisms that pose health hazards or may be useful in research. Specific gene probes can be employed, for example, to detect human pathogens in seafood and recreational waters; fish pathogens in aquaculture systems; microorganisms capable of mediating desired chemical transformations (e.g., toxic chemical degradation, CO_2 assimilation, metal reduction); and specific fish stocks in fish migration and recruitment studies.

Biomass Production

Approximately 40 percent of all primary energy production, or photosynthesis, occurs in the seas. In this process, oceanic plants (phytoplankton, seaweeds, seagrasses) take up carbon dioxide (CO_2) and, with light energy from the sun, convert it into organic carbon (primarily sugars) and oxygen. The oceans contain 50 times as much carbon dioxide as does the atmosphere, and it is estimated that primary production incorporates 35 gigatons (1 gigaton = 1 x 1015 grams) of carbon into marine biomass annually. This abundant source of fuel for energy production has not been tapped commercially because it is not competitive with soybean meal and other easily harvested, traditional sources of biomass, and also because, regardless of the source, biomass is not competitive with other types of fuels.

Research on the use of biotechnology to enhance biomass production and utility is another scope of marine biotechnology. At least three general approaches are being explored.

— The enzyme that captures CO_2 for photosynthesis-ribulose bisphosphate carboxylase\oxygenase or

"RUBISCO"-is relatively inefficient, so supercomputers are being used to verify structural information, and the enzyme is being redesigned to optimise its function.

— The chemical composition of biomass can be altered to make it more suitable for particular applications. For example, marine microalgae are being genetically engineered to boost their lipid content, with the aim of providing a source of alternative fuels that is more economical than are conventional sources.

— Biotechnology is being used to convert biomass to ethanol and other alternative forms of energy and chemical feedstocks.

Bioremediation

Bioremediation shows great promise for addressing problems in marine environments and in aquaculture. These problems include catastrophic spills of oil in harbors and shipping lanes and around oil platforms; movement of toxic chemicals from land, through estuaries, into the coastal oceans; disposal of sewage sludge, bilge waste, and chemical process wastes; reclamation of minerals, such as manganese; and management of aquaculture and seafood processing waste. The full potential for marine organisms and processes to contribute new waste treatment and site remediation technologies cannot be realised without enhanced understanding of the unique conditions in marine environments. For example, oxidation-reduction (redox) states can fluctuate in coastal and estuarine sediments. The impact of changing redox conditions on biodegradation of environmental contaminants must be understood before waste management and remediation strategies and predictive models can be developed for contaminated sediments.

Bioprocess Engineering

The emerging discipline of bioprocess engineering involves the application of biological science in manufacturing, to produce products such as biopharmaceuticals and natural bioactive agents. Bioprocess engineering requires an understanding of the biological system employed (such as a marine organism), isolation and purification of a product, and translation of the product into a stable, efficacious, and convenient form. An emerging area of interest is the potential of marine bacteria and fungi to produce unusual chemical structures with no parallels in terrestrial organisms.

Small-scale studies have begun to indicate the richness of marine microorganisms as sources for novel lead structures. The study of such materials is impeded, however, by the corrosive nature of seawater nonconventional fermentors. The use of marine organisms as a basis for bioprocess engineering depends on the development of salt-resistant reactors. In addition, numerous technical advances are required for progress in bioprocess engineering in general. For example, improved methods are needed for the isolation and purification of products from the producing organisms. Advances in this area most likely will come from emerging partnerships between the Federal Government and the private sector.

Algal Culture

Algae are a diverse group of plantlike organisms. Like plants, most algae use photosynthesis to convert carbon dioxide, water and some elements into biomass, secondary metabolites and oxygen. They differ from plants because algae don't have true roots, leaves and other structures typical for plants. The diversity in algae is enormous. Seaweeds (macroalgae) have the most complex multicellular anatomy of all algae. Some even have differentiated tissues and organs that resemble those we find in plants.

The similarities, however, are analogous, not homologous. Giant seaweeds known as kelps can be found in deeper waters. The stipes of these brown algae may be as long as 100 meter. Microalgae are much smaller than these (± 1 um) and can be found in fresh-and seawater. The great kelp beds of temperate coastal waters provide habitat and food for a variety of organism. The kelps, brown algae, are prodigiously productive. This alga: Macrocystis grows to a length of more than 60 m in a single season, the fastest linear growth of any organism.

Algal culture is chiefly used to generate biomass from which cells and metabolites can be isolated. Culturing often takes place in large ponds or raceways. Cultured algal cells also are used to minimise the production of high cost of rare compounds. This type of culturing usually takes place in a fermenter or a bioreactor rather than in ponds. Culture conditions are highly controlled and the products undergo significant downstream processing of products collection, purification and packaging. Currently, microalgae are the main organisms used in culture, although macroalgae yield valuable agars and agarose that are used n research and diagnostic labs.

Cell culture technology may be able to contribute to agar production methods, and the scarcity of agar - producing algae. Cell culture technology may be able to contribute to agar production. Speciality chemicals are produced from microalgal cultures. For example, a particular strain of Cholella has been developed that produces 30% more proline than other strains produce. The proline can be readily extracted from the culture for commercial use. A variety of hydrocarbons, polysaccharide and other important compounds also can be produced.

Gene transfer from other organisms can yield new algal products; for example, after genes are transferred from bacteria, the algae Chlamydomonas can produce the rare amino acid octopamine. Although algal cell culture is a high cost operation, the products that are isolated can be very profitable. For example, amino acid fetch $5 to $100/kilogram; food colouring phycobiliprotiens. Other industrial compounds, such as dihydroxyacetone, gluconic acid, hydrogen and acetic acid are produced by immobilised microalgal cells. One algal culture in a large fermenter potentially can genetrate thousands of dollars of speciality chemicals every month.

Medicinal Application

Marine organisms also make important contributions to medical research.. Sharks do not form cancerous cells, and are so being used to study the immune system; compounds in shark-cell extracts might hold the key to cancerous suppression. Other marine organisms serve as models for the study of human disease. For example, because the sea cucumber's abdominal cavity is filled with bacteria, the creature is being used to study resistance to peritonitis, and because the ice fish lacks red blood and haemoglobin, it is a model for anemia.

Marine organisms produce many metabolites, which are usually referred to as natural products. Chemists and pharmacologists are studying these natural products because of their unusual and sometimes unique chemical structures. Unlike primary and intermediate metabolites and cofactors, secondary metabolites are not essential for growth and reproduction, but most probably confer some selective advantage on the organism. The diverse compounds provide much diversity compounds provide much diversity of chemical structure and function.

Marine Microalgae

The diversity of microalgae is huge. Monodus produces unsaturated fatty acids and has a size of approximately 3.5 um. Algae are extremely important ecologically, accounting for about half the photosynthetic production of organic material on a global scale. As freshwater and marine phytoplankton and intertidal seaweeds, algae are the basis of aquatic food webs, supporting an enormous abundance and diversity of animals.

Biotechnological use of microalgae is very interesting: solar radiation is used as energy source and inorganic CO_2 as carbon source by autotrophic microorganisms. Both substrates are abundant and cheap. Also, micro-algae produce high-value compounds that cannot be obtained otherwise. Examples are photosynthetic pigments such as carotenoids that are commercially used as provitamin A, antioxidant and colorant agent. The aim of this project is the production and extraction of lipophilic vitamins from marine micro-algae. β-Carotene (pro-vitamin A) produced by Dunaliella salina is used as a model system.

Aquaculture

Aquaculture is a form of agriculture in which aquatic plants and animals are reared in artificial or semi-natural conditions. There are many forms of aquaculture. Typically, most people think of aquaculture as the cultivation of fish and shellfish for seafood markets and human consumption, but the scope of aquaculture is much broader. In some parts of the world, seaweeds are cultured for human food and for important food additives, such as agar and carrageenan that are sold worldwide. Aquaculture of fish and aquatic plants for the aquarium trade or for water gardens are an important business in this country, as is the aquaculture of bait fish to supply an ever-growing sport fish industry.

Aquaculture is also carried out to supply markets for shells and shell products. Most pearls, both freshwater and saltwater, have aquaculture origins. Additionally, some semi-aquatic plants, such as cranberries, might be considered to be produced through aqua cultural methods. Aquaculture, which long has been practiced in Asia and is increasingly popular in the United States, Europe, and South America, will benefit tremendously from the use of new molecular tools and processes. With worldwide seafood demand projected to increase 70 percent in the next 35 years, and harvests from capture fisheries stable or declining, aquaculture will have to produce seven times as much seafood as it generates now to supply global demand by 2025.

The use of modern biotechnology to intervene in the rearing process and enhance production of aquatic species holds great potential to meet this demand. The major research issues in aquaculture are similar to those for other agricultural sectors, but the knowledge base for aquaculture is comparatively meager. Development of this knowledge is a particular challenge due to the diversity of cultured aquatic species and the systems for their production. The scope and intensity of aquaculture can vary greatly depending upon the species cultured, financial resources of the aquaculturist, the degree of management required, and the amount of risk that the aquaculturist wishes to assume.

The application of biotechnology promises significant benefits to both producers and consumers of aquacultural products. The use of genetically enhanced organisms may improve production efficiency through improvements in growth rates, food conversion, disease resistance, and product quality and composition. The application of biotechnology to aquaculture also may help conserve wild species and genetic resources and provide unique models for biomedical research.

Biotechnology and aquaculture:

Enhancing Reproduction and Early Development—Induced Breeding: Biotechnology can be applied to enhance reproduction and early development of cultivated aquatic organisms. The resulting benefits could include year-round production of gametes and fry of economically valuable species and creation of new markets for specialised, genetically improved broodstock. Similarly, biotechnology may provide techniques for improving the reproductive success and survival of endangered species, thereby helping to preserve the diversity of life on Earth.

Improving Growth, Development, and Nutrition: Biotechnology can be applied to improve the growth, development, and nutrition of cultivated aquatic organisms. Specifically, the tools of biotechnology can enhance understanding of the environmental, biochemical, hormonal, and genetic bases of growth and nutrition for all life stages. Biotechnology also can be used to expand knowledge about the absorption, digestion, and assimilation of nutrients and to increase the efficiency of these processes.

Improving Health and Well-Being: Biotechnology offers substantial opportunities to improve the health and wellbeing of cultivated aquatic organisms. More than 50 diseases affect fish and shellfish cultured in the United States, causing losses of tens of millions of dollars annually. Biotechnology not only can improve the survival, growth, vigour, and well-being of cultivated stocks, but also can reduce disease transfer between cultivated and wild stocks. New products and market opportunities can be developed related to aquatic animal health and well-being.

The tools of molecular biology can provide a basic understanding of host immunity, resistance, and susceptibility to diseases and associated pathogens by furnishing information about life cycles and mechanisms of pathogenesis, antibiotic resistance, and disease transmission. Improved technologies must be developed for detecting and diagnosing pathogens and diseases and for enhancing the genetic basis of disease resistance, thereby reducing the need for antibiotics and other drugs. Potential products resulting from this research include gene therapy techniques; broodstock free of pathogens; safe, effective prophylactic agents, including immune modulators, antigens, and vaccines; safe, effective therapeutic agents; and improved systems for administering prophylactic and therapeutic agents.

Improving Quality and Value: Biotechnology can be employed to assess and improve the safety, freshness, colour, flavour, texture, taste, nutritional characteristics, and shelf life of aquacultural food products. In addition, practical technologies can be developed to detect and assay toxins, contaminants, and residues in seafood, and to reduce or eliminate contaminants. There are also opportunities to apply biotechnology in improving seafood processing. Research should be conducted to develop and improve technologies for all these applications.

Conserving Genetic Resources: The preservation and enhancement of biodiversity in natural systems is an important Federal priority. Therefore, the Federal Government should encourage and support programmes to maintain and enhance biodiversity in aquatic systems through cultivation and stocking of aquatic species. Biotechnology can be employed in two ways to conserve genetic resources of aquatic species.

Enhancing Biomedical Models: Aquaculture has important purposes other than food production. Because they often adapt to extreme environments, marine organisms can provide unique models for research on biological and physiological processes. Studies of the developmental, cellular, and molecular aspects of marine organisms as model systems will provide insights into the basis of disease mechanisms and pathogenesis in humans. By contrast, use of mammalian organisms as a basis for the development of some types of human disease models may be neither feasible nor cost effective. Progress in this arena will require that sophisticated molecular biology technologies be adapted to marine organisms, in order to enhance understanding of their biological processes.

For example, approaches for gene transfer into eggs have been developed for many terrestrial organisms, but not for most marine species. This technology is needed for analyses of gene regulatory systems and gene expression. In addition, methods need to be developed for culturing tissues from marine organisms. Cultured cell lines will provide opportunities for gene transfer and gene expression studies and enhance the usefulness of marine species as biomedical research models.

Aquaculture and Hydroponics: In the last 30 years, there have been methods developed for the cultivation of terrestrial plants, such as lettuce and tomato, whereby no soil is used and their roots are suspended into nutrient-laden water. This practice is called hydroponics; but, strictly speaking, this is not aquaculture because the plants are not normally considered aquatic or semi-aquatic. Aquaculture systems have been developed in which water containing wastes from aquacultured fish is used to culture hydroponic vegetables. This ingenious use of aquaculture waste resources is a form of integrated culture (aquaculture in concert with terrestrial aquaculture).

In other parts of the world, there are a number of very successful commercial applications of integrated culture. In much of Southeast Asia, a hardy African fish, called tilapia, is cultured along with rice in flooded fields. In China, it is common to fertilise fish ponds with wastes from farm animals, such as chickens or swine. In the bayou country of Louisiana, the integrated culture of crayfish and rice is common practice.

Seafood Safety: Food-borne illness is not limited to seafood, but is a common concern of all food industries. The recent media attention to seafood has led to an increase in public awareness and a number of misconceptions about the safety of eating seafood. Between 1973 and 1987, shellfish accounted for 2.8% of the cases of food-borne illness reported to the Centers for Disease Control (CDC), and finfish accounted for 2.2% of the cases. These statistics may seem high at first glance, but they are somewhat misleading.

For example, 37% of the cases of seafood-borne illness in the U.S. between 1977 and 1981 were attributed to ciguatera, a toxin found only in tropical and subtropical fish. An additional 37% of the cases during the same time period were attributed to scombroid poisoning, a toxin produced in the flesh of some species of fish when improperly stored at high temperatures. Therefore, the statistics reported by the CDC are skewed by illnesses, which either affect only a small geographical area, or only occur with mishandling of fish.

The incidence of illness attributed to seafood can be reduced if the public is better informed, understands the risks, and most importantly, learns to prevent seafood-borne illness. When handled properly, finfish and shellfish are as safe to eat as any other source of protein. For healthy individuals, the nutritional benefits of seafood far outweigh the safety concerns. Persons with compromised immune systems, such as those with liver disease, can also benefit from eating seafood but should follow a few precautionary measures when preparing seafood.

Transgenic Fish

Conventional aquaculture practices have raised a number of environmental issues including, but not limited to, pollution and the impacts on wild fish populations should fish escape from fish farms. The application of genetic engineering to aquaculture adds significant new dimensions to these existing concerns, especially the multiple possibilities related to gene flow and the ecological disruption associated with escapes.

One of the larger environmental concerns raised by transgenic fish is the possibility that a transgenic species raised in open water pens will escape and spread novel traits into the ecosystem by breeding with wild relatives, a biological process known as "gene flow." Gene flow between transgenic or conventionally bred fish and wild populations is an environmental concern because it may present a threat to natural biodiversity. According to one scientific model, if a transgenic fish escaped and mated with wild fish, gene flow could follow one of three scenarios:

— *Purge Scenario*: When the net fitness of a transgenicpopulation. This means evidence of the transgenic genome would persist in subsequent generations.

— *Trojan Gene Scenario*: When the net fitness of a transgenic fish is altered such that the fish has enhanced mating success but reduced adult viability (i.e. chances of surviving long enough to mate), introduction of that fish into the wild population could result in a rapid decline of the wild population.

> Essentially mating success would ensure the spread of the novel gene throughout the population, but the inability to survive would reduce the population size of subsequent generations and potentially lead to extinction.

Even if they don't breed with wild relatives, transgenic fish that escape into natural ecosystems could be an environmental nuisance by becoming an invasive species. This danger mainly arises for those transgenic fish endowed with new genes that improve such fitness traits as breeding capabilities and the ability to withstand harsh conditions. The establishment of a thriving transgenic fish population in an ecosystem where it has never existed could crowd out native fish populations. But scientists lack data on many ecosystem variables that would allow them to predict when and where a transgenic fish, even one clearly endowed with fitness traits, could likely become an invasive species.

It is important to note that developers of transgenic fish are attempting to reduce or eliminate both gene flow and invasive species risks by sterilising transgenic fish. Sterilisation is relatively easy and inexpensive but success rates are highly variable. In addition, sterilisation does not necessarily neutralise environmental risks. Academic scientists note that an escaped, sterile fish might still engage in courtship and spawning behaviour, disrupting breeding in wild populations. Waves of escaped sterile fish could also create ecological disruptions as each group is replaced by another equally strong group of transgenic sterile fish.

Problems on Transgenic Fish

Sometimes due to the improper and unsterile techniques used for the production of genetically modified fish, there might be negative consequences, which may lead to the contaminated product. So to avoid these problems, there are various control measures used for the maintenance of sterility of the product. One of them is Sterilisation.

The most important process to the minimisation of negative consequences associated with the production of genetically-modified fish is sterilisation. In conjunction with environmental protection, sterilisation also speeds up the fish's development and leads to better quality meat because it diverts energy normally used in reproduction to muscle development.

Environmental Impacts

Aquaculture provides a large production of nutritious, high-quality food. However, similar to the conventional agriculture, there are many adverse environmental impacts of aquaculture. The most important effects are ecological, and these are associated with the conversion of natural ecosystems into complex and intensively managed aquaculture ecosystems. For example, the conversion of tropical mangrove systems into aquacultural facilities used to raise prawns yields a combined loss of natural ecosystem.

In other words, the conversion has significant consequences for species in the environment, and usually causes damage to offshore ecosystems. With world's steadily growing population limiting supplies of food, water, minerals and energy, scientists believe we will have to rely on the vast resources of the sea as key to sustaining human life. Fortunately, the governments worldwide are striving for an increase in ocean commercialisation (the development, harvesting and marketing of the ocean's resources for a profit) to meet the rapidly growing needs of humanity and noticing the importance of preserving and protecting the marine environment.

Careless ocean commercialisation can have very serious side effects. One of the many effects is seafood and water contamination by ocean pollution posing health risks for both marine life and humans. Most of the current methods of ocean commercialisation are aquaculture, the farming of marine fish and plants; the conversion of saltwater to freshwater; utilisation of tidal and thermal energy; and the incineration of hazardous wastes.

Sea Conservation

Nowhere in the biosphere is biological diversity greater than in the seas and the extent of this diversity becomes increasingly evident as scientists investigate new environments. In the 1980s, for example, giant tube worms (Riftia) were recovered from areas adjacent to deep-ocean thermal vents, and novel mussels (Bathymodiolus) that farm methanotrophic bacteria on their gill tissue were discovered around methane seeps in the Gulf of Mexico. Most newly described species have been and will continue to be microorganisms, although it is clear that new marine plants and animals also await discovery.

Less than 1 percent of the extinct bacteria-marine and terrestrial-have been isolated and described. The marine environment represents a particularly fertile source for new bacteria, as evidenced by the recent discovery of unusual "cold water" archaea, 100-500 meters deep in the oceans. These archaea comprise a high percentage of the total bacterial ribosomal RNA present in seawater samples, yet they have not been isolated in pure culture and described. Pursuit of this research may provide new understanding of oceanic processes and, once these archaea are cultivated, perhaps a novel source of products.

Countless other marine bacteria also have yet to be cultured; advances in microbial culture technology will enable scientists to isolate, describe, and possibly exploit these organisms. Viruses are another newly appreciated element of marine biological diversity. Electron microscopy and PCR technology have revealed abundant numbers of viral particles in seawater samples. Specific viruses that infect species of marine phytoplankton have been cultured from coastal as well as open ocean sites. This discovery is exciting for at least two reasons

— Viral infection of higher forms of marine life almost certainly affects global ocean processes, such as photosynthesis.

— Marine viruses also will provide new materials for development of genetic and biotechnological tools that can be used to study and manipulate marine organisms. For example, marine viruses could be used to genetically engineer higher forms of marine life.

8

Industrial Applications of Biotechnology

Biochemicals

Biochemicals are chemicals produced from organic sources known as biomass. Biomass is any living matter that can be naturally and regularly replenished, including agricultural food and feed crop residues, aquatic plants, animal wastes and other waste materials. Biochemicals are sometimes called bio-based chemicals, green chemicals, and plant-based chemicals. Currently, the most popular source for biochemicals is corn. Chemicals like ethyl lactate, succinic acid, and polylactic acid are made from corn components, such as corn starch, and are used to make a variety of products, such as plastics, solvents, clothing fibres, paints, and food additives.

Petroleum and Corn

Petroleum is the oil found underground in certain parts of the world and is the most used energy source in the world. It is used to make gasoline, chemicals, plastics and drugs. Because petroleum is limited in quantity, and cannot be made once the supply has been depleted from the earth, it is said to be non-renewable. Petroleum products are also not environmentally friendly—causing air pollution when burned, and filling up landfills when disposed of. Biochemical products made from biomass help solve many of the problems associated with petrochemical products. Environmental benefits include less pollution from crude oil extraction and processing, as well as better waste management since the materials are biodegradable. And because biomass can be replenished regularly by nature, biochemical products are said to be renewable.

Different types of biochemicals are:

— *Commodity chemicals*: These include solvents, fuel additives, lubricants, surfactants, adhesives, and inks.

— *Fine chemicals*: These include enzymes, nutraceuticals, and pharmaceuticals.

— *Chemical intermediates*: These include sugars, organic acids, and other monomers.

Biochemicals are made from the processing of plant sugars. Currently, corn is being used extensively as a starting material for biochemicals. Making biochemicals involves two basic steps:

— First, organic matter, such as corn, is milled and the sugar is extracted from the raw material. The sugar is then fermented in a controlled environment, much like the fermentation of beer and wine. Some fermentation processes use genetically modified organisms that convert corn-derived sugar to a specific chemical.

— Next, a purification process is undertaken where the fermented liquid is purified for the target biochemical through methods such as distillation and electrodialysis. The biochemicals are then made into various products, such as plastics and solvents.

The role of biotechnology in the creation of biochemicals is crucial. Through fermentation technology, biotechnology enables the processing of common crops and crop wastes, such as corn kernels and corn stalks, to make products such as plastics, textiles, and ethanol. Biotechnology also allows the creation of genetically modified bacteria that can convert plant sugars into an ingredient used in clothing, packaging, and plastics. New enzymes are also being developed for breaking down plant sugars. Biotechnology also helps scientists better understand the biology of bacteria and enzymes that can be used in bioprocessing, which may enable the development of cheaper methods of biochemical production. Currently, biochemical products are often expensive to produce and are not cost-effective enough to compete with traditional products.

Research in biochemicals continues to grow rapidly. Many large multinational companies are investing time and money into producing chemicals that can be economically viable and environmentally friendly. For example, many chemical companies are teaming up with biotechnology companies to develop products based on biochemicals. Cargill Dow currently makes polylactic acid from corn kernels but has plans to switch to cheaper feedstocks, such as corn stalks, wheat straw, rice hulls, and sawdust, so that biochemicals can be more competitive in the marketplace.

Pollution due to the use of petroleum-based products may be eliminated by the use of biochemicals. Whereas petroleum-based products do not biodegrade and accumulate in waste landfills, biochemicals degrade completely and do not contribute to waste problems. As well, biochemicals are made from biomass; therefore, their production and use are sustainable.

Biopolymers and Bioplastics

Biopolymers are polymers which are present in, or created

by, living organisms. These include polymers from renewable resources that can be polymerised to create bioplastics. Bioplastics are plastics manufactured using biopolymers, and are biodegradable. Biopolymers and bioplastics are not new products. Henry Ford developed a method of manufacturing plastic car parts from soybeans in the mid-1900s. However, World War II side-tracked the production of bioplastic cars. Today, bioplastics are gaining popularity once again as new manufacturing techniques developed through biotechnology are being applied to their production.

There are two main types of biopolymers: those that come from living organisms; and, those which need to be polymerised but come from renewable resources. Both types are used in the production of bioplastics.

Biopolymers are present in, or created by, living organisms. These include carbohydrates and proteins. These can be used in the production of plastic for commercial purposes. Examples are listed in the table below.

Biopolymer	Natural Source	What is it?
Cellulose	Wood, cotton, corn, wheat, and others	This polymer is made up of glucose. It is the main component of plant cell walls.
Soy protein	Soybeans	Protein which naturally occurs in the soy plant.
Starch	Corn, potatoes, wheat, tapioca, and others	This polymer is one way carbohydrates are stored in plant tissue. It is a polymer made up of glucose. It is not found in animal tissues.
Polyesters	Bacteria	These polyesters are created through naturally

occurring chemical reactions that are carried out by certain types of bacteria.

Polymerisable Molecules

These molecules come from renewable natural resources, and can be polymerised to be used in the manufacture of biodegradable plastics. There are two methods being researched and used to produce plastics from plants. The first uses fermentation, and the second relies on the plant to become the factory for plastic production. These two methods are outlined below.

Fermentation

Fermentation, used for hundreds of years by humans, is even more powerful when coupled with new biotechnology techniques. Fermentation is the use of microorganisms to break down organic substances in the absence of oxygen. Today, fermentation can be carried out with genetically engineered microorganisms, specially designed for the conditions under which fermentation takes place, and for the specific substance that is being broken down by the microorganism. There are two ways fermentation can be used to create biopolymers and bioplastics:

Bacterial Polyester Fermentation: Bacteria are one group of microorganisms that can be used in the fermentation process. Fermentation, in fact, is the process by which bacteria can be used to create polyesters. Bacteria called Ralstonia eutropha are used to do this. The bacteria use the sugar of harvested plants, such as corn, to fuel their cellular processes. The by-product of these cellular processes is the polymer. The polymers are then separated from the bacterial cells.

Lactic Acid Fermentation: Lactic acid is fermented from sugar, much like the process used to directly manufacture polymers by bacteria. However, in this fermentation process, the final product of fermentation is lactic acid, rather than a polymer. After the lactic acid is produced, it is converted to polylactic acid using traditional polymerisation processes.

Plastics in Plants

Plants are becoming factories for the production of plastics. Researchers created a Arabidopis thaliana plant through genetic engineering. The plant contains the enzymes used by bacteria to create plastics. Bacteria create the plastic through the conversion of sunlight into energy. The researchers have transferred the gene that codes for this enzyme into the plant, as a result the plant produces plastic through its cellular processes. The plant is harvested and the plastic is extracted from it using a solvent. The liquid resulting from this process is distilled to separate the solvent from the plastic.

Biotechnology is driving the production of new bioplastics. Biotechnology techniques used to produce bioplastics include fermentation, and genetic engineering. For example, fermentation is used to release the cellulose from plants, so the cellulose can be used to create plastics. Also, genetic engineering can be used to create plants, such as soybean, specifically designed to be used as a raw material for the production of bioplastics.

Improving efficiency is a major concern for the production of plastics and bioplastics. Currently, fossil fuel is still used as an energy source during the production process. This has raised questions by some regarding how much fossil fuel is actually saved by manufacturing bioplastics. Only a few processes have emerged that actually use less energy in the production process.

Therefore, researchers are still working on refining the processes used in order to make bioplastics viable alternatives to petrochemical plastics. Energy use is not the only concern when it comes to biopolymers and bioplastics.

There are also concerns about how to balance the need to grow plants for food, and the need to grow plants for use as raw materials. Agricultural space needs to be shared. Researchers are looking into creating a plant that can be used for food, but also as feedstock for plastic production. One group is attempting to genetically engineer corn to contain the bacterial enzyme responsible for plastic production. Eventually, they are hoping to create the plant in a way which would restrict the plastic production to the stem, and leaves of the plant. This would leave the edible part of the corn plastic free. The edible part of the corn would be used as food, or as livestock feed. The plastic would be removed from the remaining part of the corn plant.

Biopolymers and bioplastics are the main components in creating a sustainable plastics industry. These products reduce the dependence on non-renewable fossil fuels, and are easily biodegradable. Together, this greatly limits the environmental impacts of plastic use and manufacture. Also, characteristics such as being biodegradable make plastics more acceptable for long term use by society. It is likely that in the long term, these products will mean plastics will remain affordable, even as fossil fuel reserves diminish.

Textiles Industry

Biotechnology has impacted the textiles industry through the development of more efficient and more environmentally friendly manufacturing processes, as well as through the design of improved textile materials. Some of biotechnology's key roles have involved the implementation, production, and modification of enzymes for the improvement of textile manufacturing processes. Biotechnology has also facilitated the production of novel and biodegradeable fibres from biomass feedstocks.

Enzyme Biotechnology

Through biotechnology, enzymes are used to treat and modify fibres during textile manufacturing, processing, and in caring for the product afterwards. Some applications include:

De-sizing of cotton: Untreated cotton threads can break easily when being woven into fabrics. To prevent this breakage, they are coated with a jelly-like substance through a process called sizing. However, after the threads have been woven into fabrics, the agents needed to further finish the material cannot adhere to the jelly-coated fabrics. Thus, the protective sizing agents must be removed by a process called de-sizing. Amylase enzymes are widely used in de-sizing, as they do not weaken or affect cotton fibres, nor do they harm the environment.

Retting of flax: Flax plants are an important source of textile fibres. Useful flax fibres are separated from the plant's tough stems through a process called retting. Traditional retting methods consume large quantities of water and energy. Bacteria, which may be bred or genetically engineered to contain necessary enzymes, can be used to make this a more energy efficient process.

Breakdown of hydrogen peroxide: When cotton is bleached, a chemical called hydrogen peroxide, which can react with other dyes, remains on the fabric. Catalase enzymes specifically break down hydrogen peroxide and may be used to remove this reactive chemical before further dyeing.

Biostoning and Biopolishing: Instead of using abrasive tools like pumice stones to create a stonewashed effect or to remove surface fuzz, cellulase enzymes may be used to effectively stonewash and polish fabrics without abrasively damaging the fibres.

Detergents: Enzymes allows detergents to effectively clean clothes and remove stains. They can remove certain stains, such as those made by grass and sweat, more effectively than enzyme-free detergents. Without enzymes, a lot of energy would be required to create the high temperatures and vigorous shaking needed to clean clothes effectively. Enzymes used in laundry detergents must be inexpensive, stable, and safe to use. Currently, only protease and amylase enzymes are incorporated into detergents. Lipase enzymes break down too easily in washing machines to be very useful in detergents. However, their stability is being studied and further developed through methods such as genetic screening and modification.

Novel Fibres

Synthetic fibres made from renewable sources of biomass are environmentally sustainable, and are becoming increasingly economically sustainable. Biodegradable synthetic polymers include novel fibres such as polyglycolic acid and polylactic acid, which are made from natural starting materials. Not all novel fibres are synthetic; they may also be naturally derived. Some natural biological fibres come from basic materials found in nature, including:

— Chitin: a type of sugar polymer found in crustaceans

— Collagen: a type of protein found in animal connective tissue

— Alginate: a type of sugar polymer found in certain bacteria

A prime example of a synthetic biomass fibre is Polylactic Acid (PLA), which is made by fermenting cornstarch or glucose into lactic acid, and then chemically transforming it into a polymer fibre. With properties similar to other synthetic fibres, PLA based materials are durable with a silky feel, and may be blended with wool or cotton.

Polylactic Acid (PLA)

PLA has potential applications in several areas, including the following:

— *Textiles*: clothing, fashions, upholstery

— *Agriculture*: plant mats, tree nets, soil erosion control products

— *Sanitation*: household wipes, diaper products

— *Medicine*: disposable garments, medical textiles

PLA minimises environmental waste, as it may be fully biodegraded by microorganisms under appropriate conditions into carbon dioxide and water. Unlike the non-renewable petroleum resources used to make traditional synthetic fibres, the supply of renewable corn biomass needed to make PLA is expected to surpass demand in the anticipated future.

Medical Textiles

Biodegradable synthetic fibres and natural biological fibres may be used to make textiles for medical applications. Such textiles may be used in first aid, clinical, and hygienic practices. Some examples are described below:

Polymer	Use(s)
Polylactic Acid and Polyglycolic Acid	Used in sutures, absorbable wound closure products, orthopaedic repair absorbable pins, and fixation devices, as well as in tissue engineering structures

Chitin	Incorporated into wound dressings
Collagen	Uses in cell engineering structures, such as in artificial skin, or even as surgeon's thread
Alginate	Used to protect and interact with wounds

Synthetic biomass fibres may also be used in drug delivery systems, which are designed to release drugs at a specified rate for a specified time.

ENZYMES

Enzymes are proteins that speed up biochemical reactions without being consumed or changed by the reaction. They are found throughout nature; in our bodies, in the environment, and in all living things. Without enzymes, life would not be possible. Enzymes speed up chemical and biochemical reactions, and this process is called catalysis. The substances upon which enzymes work when performing catalysis are known as substrates, and each enzyme will only fit and act upon a specific set of substrates.

There are so many enzymes that it would be impossible to name them all. In fact, scientists have yet to discover many enzymes, or fully understand their structure and properties. On the other hand, many other enzymes have been successfully studied and applied to industrial and commercial uses. A few basic enzyme types are briefly described as follows:

Type	Enzyme Function
Cellulase	Breaks down cellulose, a fibre found in the cell walls of all plants and trees. Cellulose is the basic raw material used to make products such as paper, cotton, and other textiles
Hemicellulase	Breaks down hemicellulose, another plant sugar that is not as complex as cellulose and is easier to break down

Xylanase	Breaks down xylan, a gummy sugar present in the cell walls of plants and trees. This enzyme type is used primarily in the wood and pulp industry
Amylase	Breaks down starches and other carbohydrates into basic sugars
Protease	Breaks down proteins
Lipases	Breaks down fats

In the absence of enzymes, chemical reactions occur only when molecules collide while in proper alignment with each other. Because molecules are bumping into each other randomly, chemical reactions are essentially due to chance events. This sometimes results in reactions that occur very slowly, or reactions that do not occur at all. Enzymes act like tiny molecular machines to ensure that molecules come into contact with each other and react.

Like a key fitting into a lock, chemical molecules fit into pocket-like structures located on an enzyme. These pockets hold the molecules in a position that will allow them to react with each other, ensuring that they are close enough together and aligned properly for a reaction to occur. In this way, enzymes speed up reactions. The enzymes are not changed themselves by the reaction. When the reaction is complete, enzymes release the product(s) and are ready to bring together more molecules and catalyse more reactions.

Biotechnology can provide an unlimited and pure source of enzymes as an alternative to the harsh chemicals traditionally used in industry for accelerating chemical reactions. Enzymes are found in naturally occurring microorganisms, such as bacteria, fungi, and yeast, all of which may or may not be genetically modified. Large quantities of enzymes are often needed for industrial use, so these microorganisms are multiplied through a process called fermentation. When enzymes or the microorganisms used to create them are no longer needed, they are destroyed through exposure to heat or safe organic/ inorganic materials.

Pulp and Paper Industry

Enzymes have been used in the pulp and paper industry to soften wood fibres, improve drainage, and present alternatives to chemical bleaching.

Biopulping: Paper is made from cellulose fibres, which must be separated from a tough wood fibre called lignin. The step by step process used to separate cellulose from lignin and other wood components is known as pulping. It is a time and energy consuming process, involving the mechanical processing of wood or the treatment of wood with harsh chemicals. In biopulping, cellulase and xylanase enzymes made by lignin-degrading fungi are used to pre-treat wood and break down the lignin fibres. Removing lignin prior to further wood pulping saves time and energy, and decreases the quantities of chemicals used.

Draining: Enzymes can also improve water drainage during wood pulping, a process that often slows down paper production. When fine lignin fibres are degraded by enzymes, less water is absorbed, thereby reducing drainage times, lessening the energy required to dry the paper, and producing a cleaner water runoff.

Bleach Boosting: Lignin fibres that remain in wood pulp are coloured and must be bleached, usually by harsh chlorine compounds under high pressures. As an alternative, enzymes may be used to remove fine surface fibres, thereby reducing the bleaching process or eliminating it altogether.

Enzyme Biotechnology and Textiles

Enzymes are used to treat and modify fibres, particularly

during textile processing and in caring for textiles afterwards. For example, enzymes called catalases are used to treat cotton fibres and prepare them for the dyeing processes. Some bacterial enzymes are used to separate the tough stem of the flax plant from the flax fibres used in textiles. By degrading surface fibres, many enzymes, including some cellulases and xylanases, are used to finish fabrics, give jeans a stonewashed effect, or help in the tanning of leathers. A recombinant enzyme called laccase, made by certain fungi, may also be used to treat fabrics and even catalyse the synthesis of some synthetic fibres.

There are even enzymes in regular laundry detergent to help break down dirt, clean clothes more effectively, and prevent the dulling of fabric colours. Enzymes are frequently used in laundry detergents. Without them, very high temperatures and mechanical shaking would be required to effectively clean clothes and other textiles. The main enzyme types used in laundry detergents are briefly described as follows:

Enzyme Type	Example of Sources	Brief Description
Protease	Bacillus licheniformis B. amyloliquefaciens	Breaks down protein-based dirt. May be made resistant to oxidation through protein engineering.
Amylase	Bacillus licheniformis	Breaks down starch-based dirt. May be made resistant to oxidation through protein engineering.
Lipase	Aspergillus oryzae	Breaks down fats and oils. Most are not sufficiently stable in washing machines to be very useful. Improved enzymes are being developed through

		genetic screening and modification.
Other	Various	Examples include: Lipoxygenase and glucose oxidase enzymes, which produce hydrogen peroxide. Cellulases, which degrade surface fibres to prevent colour fading and fuzzing.

Biofuels

Enzymes may be used to help produce fuels from renewable sources of biomass. Such enzymes include cellulases, which convert cellulose fibres from feedstocks like corn into sugars. These sugars are subsequently fermented into ethanol by microorganisms. A new process called Simultaneous Saccharification and Fermentation has greatly improved ethanol production efficiency. In this new process, cellulase enzymes and fermentation microorganisms are combined in a single reaction mixture to produce ethanol in one step, rather than producing sugars from cellulose and then fermenting them into ethanol separately.

Advances in science and technology have allowed researchers to improve enzymes through modifications to enzyme producing microorganisms, or through direct changes to the enzymes themselves. Genetic Engineering and Recombinant DNA technology—By using recombinant DNA technology, microorganisms may be genetically modified to produce a desired enzyme under specific conditions. This is accomplished through recombinant DNA technology, whereby small circular pieces of DNA, known as plasmids, are used to insert enzyme-producing genes into the genomes of organisms that possess another desirable trait, such as the ability to thrive on inexpensive nutrients. Therefore, both the enzyme and the original trait will be expressed in a single recombinant microorganism.

Protein Engineering: All enzymes are made of proteins, which are large molecules formed from basic units, called amino acids, strung together like beads on a chain. To form functional enzymes, long chains of amino acids must be folded properly. Some enzymes consist of only one chain, whereas others are made of several chains that fit together. Scientists are using technology to study how proteins are formed, how they fold, and how they function. By studying the relation between the structure of a protein and how it functions, they are developing ways to improve and engineer enzymes. Proteins may be modified by changing one or more amino acids, and/or changing the way the amino acid chains fold and fit together.

Enzymes are a sustainable alternative to the use of harsh chemicals in industry. Because enzymes work under moderate conditions, such as warm temperatures and neutral pH, they reduce energy consumption by eliminating the need to maintain extreme environments, as required by many chemically catalysed reactions. Reducing energy consumption leads to decreased greenhouse gas emissions by power stations. Enzymes also reduce water consumption and chemical waste production during manufacturing processes. Because enzymes react specifically and minimise the production of by-products, they offer minimal risk to humans, wildlife, and the environment. Enzymes are both economically and environmentally feasible because they are safely inactivated and create little or no waste; rather than being discarded, end-product enzymatic material may be treated and used as fertiliser for farmers' crops.

Microbial Enhanced Oil Recovery (MEOR)

Microbial Enhanced Oil Recovery is the use of microorganisms to retrieve additional oil from existing wells, thereby enhancing the petroleum production of an oil reservoir. In this technique, selected natural microorganisms are introduced into oil wells to produce harmless by-products, such as slippery natural substances or gases, all of which help propel oil out of the well. Because these processes help to mobilise the oil and facilitate oil flow, they allow a greater amount to be recovered from the well.

MEOR is used in the third phase of oil recovery from a well, known as tertiary oil recovery. Recovering oil usually requires two to three stages, which are briefly described as follows:

1. *Primary Recovery*: 12% to 15% of the oil in the well is recovered without the need to introduce other substances into the well.
2. *Secondary Recovery*: The oil well is flooded with water or other substances to drive out an additional 15% to 20% more oil from the well.
3. *Tertiary Recovery*: This stage may be accomplished through several different methods, including MEOR, to additionally recover up to 11% more oil from the well.

The microorganisms used in MEOR can be applied to a single oil well or to an entire oil reservoir. They need certain conditions to survive, so nutrients and oxygen are often introduced into the well at the same time. MEOR also requires that water be present. Microorganisms grow between the oil and the well's rock surface to enhance oil recovery by the following methods:

Reduction of oil viscosity: Oil is a thick fluid that is quite viscous, meaning that it does not flow easily. Microorganisms help break down the molecular structure of crude oil, making it more fluid and easier to recover from the well.

Production of carbon dioxide gas: As a by-product of metabolism, microorganisms produce carbon dioxide gas. Over time, this gas accumulates and displaces the oil in the well, driving it up and out of the ground.

Production of biomass: When microorganisms metabolise the nutrients they need for survival, they produce organic biomass as a by-product. This biomass accumulates between the oil and the rock surface of the well, physically displacing the oil and making it easier to recover from the well.

Selective plugging: Some microorganisms secrete slimy substances called exopolysaccharides to protect themselves from drying out or falling prey to other organisms. This substance helps bacteria plug the pores found in the rocks of the well so that oil may move past rock surfaces more easily. Blocking rock pores to facilitate the movement of oil is known as selective plugging.

Production of biosurfactants: Microorganisms produce slippery substances called surfactants as they breakdown oil. Because they are naturally produced by biological microorganisms, they are referred to as biosurfactants. Biosurfactants act like slippery detergents, helping the oil move more freely away from rocks and crevices so that it may travel more easily out of the well.

MEOR is a direct application of biotechnology. It uses biological materials, such as bacteria, microorganisms, and their products of metabolism to facilitate the movement of oil out of a well, thereby enhancing oil recovery. Other applications of biotechnology in MEOR include genetic engineering techniques and recombinant DNA technology, which are used to develop strains of bacteria with improved oil recovery traits. By inserting genes from one type of bacteria into another, scientists may combine two desirable genetic traits into one microorganism.

For example, the temperature within an oil well is often too high for most microorganisms to survive. By inserting a gene that codes for a bacteria's ability to aid oil recovery into the genome of an existing bacteria that can survive under high temperatures, scientists may produce microorganisms that can both survive the heat of an oil well and also help retrieve oil. On their own, each bacteria lacks a trait necessary for oil recovery operations, but when combined through genetic engineering, the bacteria become integral to MEOR.

The environmental conditions in an oil well make it very difficult for bacteria to survive, and those that do often have a decreased ability to carry out the chemical processes needed to enhance oil recovery. Researchers are working to create strains of bacteria that are better able to survive such harsh conditions but still retain the ability to carry out the chemistry needed for MEOR. Genetic engineering is being used to develop microorganisms that can not only live in the high temperatures of an oil well, but can also subsist on inexpensive nutrients, remain chemically active, and produce substantial amounts of biosurfactants. Some researchers are developing bacteria that can be grown on inexpensive agricultural waste material, which is abundant in supply and is environmentally friendly.

As MEOR reduces or eliminates the need to use harsh chemicals during oil drilling, it is an environmentally compatible method of carrying out tertiary oil recovery. MEOR will become increasingly economically feasible as genetic engineering develops more effective microbial bacteria that may subsist on inexpensive and abundant nutrients. Methods for developing and growing MEOR bacteria are improving, thereby lowering production costs and making it a more attractive alternative to traditional chemical methods of tertiary oil recovery.

Starch Bioprocessing Techniques

Bioprocessing is the use of biological means, such as the action of enzymes or other microorganisms, to treat various materials using a set of defined procedures. When this process is used to treat starch-based biomass, it is referred to as starch bioprocessing. Starch is a complex carbohydrate, found in cereal crops like wheat and corn. Because starch-containing biomass forms the basis of many lubricants, fuel additives, and other biochemicals commonly sold on the market, starch bioprocessing is often used in product manufacturing, in the conversion of waste materials to useful by-products, and in the breakdown of waste materials.

Raw biomass from various organic sources, including wheat and corn, is the starting material used to make lubricants, fuel additives, and other biochemical products through bioprocessing. Because these products are commonly sold on the market, they are known as commodity biochemicals. The following steps outline the production of commodity biochemicals from a corn-based source of biomass:

1. *Pretreatment*: A process called corn wet milling is used to clean and treat corn raw materials before they are used as feedstocks in bioprocessing. Corn kernels are cleaned, soaked in a warm acidic liquid, and milled to produce a pulpy substance composed of corn starch, protein, and oil. Pretreatment requires little energy, and all products of this step are used. For instance,

much of the pulpy substance is made into sweeteners, such as glucose and corn syrup, while the corn starch is reserved to make commodity biochemicals.

2. *Hydrolysis and Bioprocessing*: Hydrolysis is a general term used to describe a chemical reaction that uses water to break down a larger compound, such as starch, into two or more smaller compounds, like sugars. Traditionally, harsh acids were used to initiate or speed up hydrolysis reactions, but in starch bioprocessing, such acids have been replaced by environmentally friendly microorganisms or enzymes. Bacterial enzymes are well suited for industrial use since they are inexpensive to produce and are stable at various temperatures. In this manner, enzymes bioprocess starches into sugars, which can then be made into commodity biochemicals.

3. *Further Bioprocessing*: By studying how microorganisms produce particular biochemicals, scientists will be able to identify the genes that control these functions. It may be possible, then, to use recombinant DNA technology or cell fusion to introduce such genes into microorganisms and genetically engineer them to process biochemicals more effectively and efficiently. By creating better microorganisms, starch bioprocessing techniques may be improved.

Corn starch is made of a long repeating chain of many small sugar molecules. During a hydrolysis reaction, the links between these molecules are broken down, resulting in single units of sugar. These single units are referred to as dextrose, or corn sugar.

The use of biotechnology is essential in starch bioprocessing techniques. Such biotechnologies use microorganisms, enzymes, or other biological materials to break down starches into sugars and further process them into chemicals. Biotechnology may also be used to genetically modify these microorganisms to increase the efficiency of sugar and chemical production from starches. By studying how microorganisms produce particular biochemicals, scientists will be able to identify the genes that control these functions. It may be possible, then, to use recombinant DNA technology or cell fusion to introduce such genes into microorganisms and genetically engineer them to process biochemicals more effectively and efficiently. By creating better microorganisms, starch bioprocessing techniques may be improved.

Using microorganisms and enzymes in biochemical processing increases efficiency and decreases waste production. Biological techniques and other renewable processes, such as the use of enzymes to convert biomass resources into commodity biochemicals, must be increasingly used if biochemical production methods are to be environmentally sustainable.

Future of Industrial Biotechnology

In the USA in 2001, the Biomass R&D Board had surveyed 134 industrial sites that were manufacturing bioproducts and found 440 of which produced bioenergy. In 2001, about 5 per cent of global chemical sales were derived in part form industrial biotechnology. That figure could reach 20 per cent by 2010, if chemical firms, whose polymer innovation had stalled in recent decades, became convinced by the economics of bio-based products.

Enzymes and fermentations were already used in the production of flavours and fragrances, while other markets would still be dominated by conventional chemistry through 2010 and beyond. The first applications in the largest volume segments—polymers and bulk chemicals—have already been commercialised. However, in these largely costdriven segments, a number of technological advances and policy measures would determine the ultimate uptake of industrial biotechnology.

As biotechnological processes become cheaper, the economic contribution of industrial biotechnology will increase. Among the incentives for industrialists to invest in bioproducts or biofuels, which were still more expensive than their chemical equivalents, the OECD suggested to include the real cost of the environmental impact of industrial processes in market prices.

The American Institute of Chemical Engineers proposed a total cost assessment method, which attempted to figure out in dollars all the risks posed on the environment, as well as on health and safety. Such a method and others like the 'green index' should entice an increasing number of enterprises to seize the opportunities offered by the bio-industry not only for the environment but also for the consumers and shareholders.

In the USA, representatives from different governmental bodies, industry, agriculture and academia worked together on a project called 'Vision 2020' with a view to boosting industrial biotechnology over the next decades backed by low feedstock prices and the support of the agricultural community in the USA. The Farm Bill, passed in May 2002 by the US House of Representatives, authorised $5 million in 2002 and $14 million a year from 2003 through 2007 to fund biomass research and award grants to build 'biorefineries'—factories converting biomass into chemicals, fuels and energy. The Bill also requested the government to give preference to purchasing bio-based products.

Meanwhile, the European Bioindustries Association (EuropaBio) has proposed to the European Commission to form a strategic alliance to design a so-called Technology Platform. In order to stimulate global companies to invest in industrial biotechnology and in Europe, bridging incentives, more effective regulatory processes and low-cost feedstock were key factors. Another switch that may prove crucial to watch is the move from ethanol to biotechnologically produced hydrogen.

The driving power would not come from internal combustion engines but from fuel cells using hydrogen. In this regard, genomics may be the answer. Using the money Celera Genomics had raised, Craig Venter, who designed the whole-genome shotgun sequencing for both the human and microbial genomes, set up the Institute for Biological Energy Alternatives, with a view to studying microorganisms that produce hydrogen. One example is that of Carboxydothermus, discovered in a hydrothermal vent (deep-sea volcanic spring) off the coast of Russia, which draws its energy from reacting carbon monoxide with water, and produces hydrogen as a waste product.

Another of Craig Venter's projects was to create a bacterium with a completely synthetic genome, leaving out the genes for carbohydrate synthesis that normally use hydrogen ions, so that the bacterium could devote all its energy pathways to producing hydrogen. Such unicellular micro-organism would not be able to live outside the laboratory because it would lack the biochemical mechanisms to survive there. Discovering new materials for the industry while drawing on natural resources also speaks well for the prospects of industrial biotechnology.

For instance, Nexia Biotechnologies, based in Quebec, is using technology similar to that of GTC Biotherapeutics to manufacture spider silk in goat's milk. Spider silk has been produced for the past 400 million years and has been adapted to a wide range of uses. This is an opportunity human beings could seize and use biotechnology-derived spider silk in many applications, from surgery to the design of new textiles.

9

Applications of Biotechnology in Health Care

Biotechnology which has emerged as harbinger of new drug development in the 1990's and this decade can be called as the decade of Biotechnology. The advent of genetic engineering and hybridioma technology has led to the production of many newer drugs such as genetically engineered proteins and peptides which include hormones, enzymes, interferon's, vaccines and monoclonal antibodies. The different classes of drugs obtained from biotechnology are:

— Anticoagulants/ Thrombolytics agents Prourokinase for the treatment of heart attacks.

— Colony stimulating factors (CSF) Granulocyte CSF, for the treatment of AIDS, leukemia and a plastic anemia, and granulocyte Macrophage CSF, for same indications as above and Hodgkin's disease.

— Erythropoietins to treat anemia

— Interferons several types are under study to treat cancer, AIDS, herpes and rheumatoid arthritis.

— Interlocking Interleukin-2, for cancer immunotherapy.

— Monoclonal Antibodies For variety of cancer conditions as well as in the treatment of septic shock and to control heart and liver transplant rejection.

— Peptides Endogenous peptides are under investigation to treat autoimmune diseases and congestive heart failure.

— Vaccines To treat hepatitis B, Hemophilus influenza type B, Cancer, Malaria and AIDS.

According to a survey made by PhRMA (Pharmaceuticals Research and Manufacturers of America) in 1995, the number of biotechnological drugs under clinical trials has grown by 64 percent in 1.5-year time. Till now, 24 drugs have been approved by American FDA (Food and drug Administration). There are about 1100 biotechnology companies in America. According 1991 prediction, the US pharmacy Market is supposed to generate about one billion US$ and also the market is estimated to grow at a rate of 25 percent per year Throughout this decade. In this millennium, the sales are predicted to reach 11 billion US$. Clearly biotechnology has made a significant contribution to the health care sector over the last decade and the same trend is supposed to continue in future.

Hormones

In this class, the first product to get FDA approval was Insulin. The other hormones are human growth hormones.

Insulin: It is first product from genetic engineering to get approval by FDA. The production of human insulin is done by two methods, both using rDNA technology. In the first method, the two chains of insulin (α and β) are produced by separate fermentations and were joined together. In the second method, proinsulin is prepared, which has a third c-peptide, which hold alpha and beta chains together. Enzymatic cleavage of c-peptide results in the active insulin. Multicenter clinical studies have shown no difference between the two products.

Growth hormone: Up to 1995, the source of growth hormone was the pituitary gland. In 1985, a genetically engineered analogue of human growth hormone (hGH) came to market (company, Genentech). It was called a Somaterm. This has an extra methionine at the end of molecule. This drug got FDA approval in 1985. In 1987, another drug somatropin (Humatrope) manufactured by Eli Lilly got approval. This was identical with growth hormone from pituitary. Genetech has forwarded an application to FDA for the use of protropin in retarded growth and sexual development, a condition called as Turner's syndrome.

ANTIBODIES

The first antibody-derived product to get approval by FDA was digoxin immune Fab (Digibind) manufactured by Burroughs welcome, in 1986. This product is for utilisation in digitalis toxicity. An antigen is created by conjugating digoxin molecule to human serum albumin. The conjugate is injected to sheep to produce digoxin specific antibodies. The antiserum of sheep with high affinity and specificity for digoxin specific Fab fragments. These are purified by affinity chromatography. Hybridoma technology is used to produce monoclonal murine antibodies. When injected into a digitalis toxic patient, these Fab fragment bind to digoxin and this complex is excreted through kidneys.

The treatment gives no adverse effects. The first therapeutic application of MAbs was in development of immunosuppressant drug muromonab, CD3 (OKT3). This is used for treatment of renal transplant rejection. OKT3 is murine antibody directed to a glycoprotein on the surface of T cells, which is essential for T cell function. This glycoprotein has 20000 molecular weight and is called CD3. It is associated with antigen recognition structure of T cell and is essential for signal transduction. In a study OKT3 reversed 94 percent of rejection, whereas therapy with steroids demonstrated only 75 percent reversal. MAbs are the leading category of all the biotechnological products. But till now, only one product has been approved.

VACCINES

In 1986 hepatitis B vaccine (Recombivax HB-Merck, Rahway) received FDA approval this was the first vaccine developed by recombinant DNA technology. Before 1986, the vaccine was obtained from plasma of healthy HbsAg (surface antigen of hepatitis B virus) carriers. But the genetically engineered product was obtained by cloning of gene responsible for HbsAg into fungi, Saccharomyces cervisiae. After the expression of protein, it is purified, absorbed into aluminum hydroxide and can be used as vaccine.

In 1989, SmithKline Beecham received FDA approval for recombinant version of the vaccine Engerix B, which is similar to recombivax HB in all aspect-dosing regimen. In 1988, Haemophilus influenza type B vaccine (Hibtiter proxis biological) got approval. The original vaccine was derived from polyribosyl ribitol phosphate (PRP), a capsular polysaccharide from the bacteria. The problem with this vaccine was its limited immunogenecity in children below 2 years. The immunogenecity was improved by conjugating the vaccine with the diphtheria toxoid protein, which act as a carrier. The combination has shown a good response.

THROMBOLYTICS

The first thrombolytic to enter into market is Activase (Genetech), a recombinant tissue plasminogen activator (TPA). It is indicated for the treatment of myocardial infraction. But this product has no cost benefit ratio when compared to other Thrombolytics. All Thrombolytics act by the activation of plasminogen to plasmin, which degrades fibrin clots, as shown below.

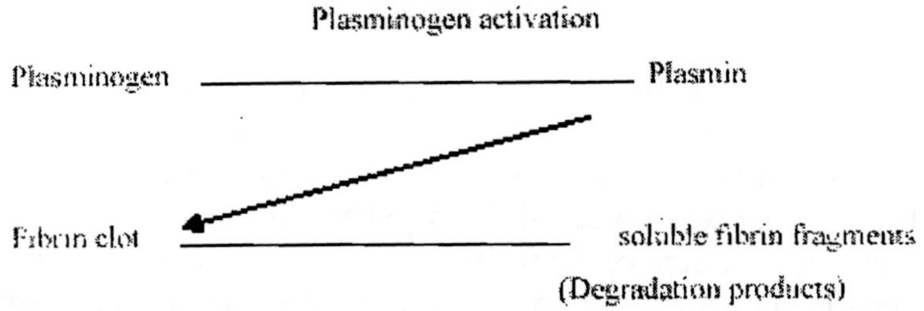

This is also degrades fibrin located throughout systemic circulation. This leads to excessive bleeding. Thus attempts are being made to produce a clot specific agent, which will not produce plasmin activity outside the clot many tissues produce proteases, which activate plasminogen. The gene that codes for such tissue plasminogen activator can be cloned into vector and expressed to produce TPA. The drug had clot specificity in vitro, but it was not found when given to humans.

TPA has been compared to many Thrombolytics. One such study compared TPA to streptokinase (SK), produced by β-hemolytic streptococci. TPA cleaves plasminogen directly to plasmin whereas SK forms 1:1 complex with plasminogen to plasmin. The production cost of SK is very low. Hence the advantage of TPA is to be determined. Another thrombolytic agent introduced in the market is an isolated plasminogen activator complex (APSAC). This product gives a thrombolytic activity by timed release supply due to complex dissociation.

It is compared to TPA and SK. All these have shown to be equally effective in mortality reduction. TPA is only genetically engineered product and it has also received approval for treatment of pulmonary embolism. There is another recombinant DNA product, factor VIII, for the treatment of clotting disorder in hemophilic patient. These patients suffer internal bleeding due to deficiency of factor VIII the treatment can be given by infusion of this protein derived from human blood. But by this method, there are chances of getting HIV infection. The rDNA product is 100 percent HIV free and has greater purity. The gene for factor VIII is isolated and cloned in mammalian cells to get the protein, which can be purified by affinity chromatography. The FDA is reviewing this product for approval.

Tumor Necrosis Factors

These substances serve as intercellular messengers to trigger an immune response against tumor cell. The blood supply to tumor cells is cut off, blood vessel to the tumor cells are destroyed and they are deprived of nutrition. Many products of this category are under early clinical stage.

DNAase

The recombinant human DNAase got approval in 1993. This is used in treatment of cystic fibrosis, a lung disease characterised by viscous purulent secretions. Large amount of DNAase responsible for this condition Bovine pancreatic DNAase was used to treat this condition. Genetech has produced human DNAase by recombinant DNA technology. This product has shown in much improvement in lung function after inhalation.

Lymphokines

Interferons are the first substances in the class of lymphokines and cytokines to receive FDA approval in 1986, a-interferons came to market. These are the protein, which act as messengers between the cells to help coordinate an immune response against foreign substances. In cancer the communication between the cells is cut off and in such situation cytokines help to establish an immune response against tumor cells. The interferon's are of basically three types namely α,β and γ. Within the a class there are at least 16 subtypes.

Two important types that got FDA approval are α-2a and α-2b. Interferon's α-2a is manufactured by Hoffmann laRoche (Referon-A) and interferon α-2b is by Schering-plough (Intron-A). These two are indicated for the treatment of hairy cell leukemia. Intron A and Alferon N (Interferon α-n3) are approved for use in the treatment of genital warts (in 1988 and 1989) Intron A and Referon A are approved for the treatment of AIDS related Kaposi's sarcoma in 1988. Interon A is also approved for the treatment of chronic hepatitis B and non-A and non-B hepatitis.

Application is also submitted for the use of Intron A in basal cell carcinoma and superficial bladder cancer. In 1990, Actimmune (Interferon-1b, Genetech) got approval for reducing the frequency and severity of infections associated with chronic glaucomatous disease in a study among 63 patients treated with interferon, 14 had serious infections whereas among 65 patients treated with placebo, 30 got the disease. Phase III clinical studies are going on for utilisation of act immune in small cell lung cancer and atopic dermatitis.

There are large numbers of lymphokines and related products under development, which include interferon's, interleukins and colony stimulating factors (CSF). The two major types of colony stimulating factors are granulocyte monocyte CSF (GM-CSF) and granulocyte CSF (G-CSF). There are some differences between the two types both these have different clinical profile. GM-CSF has shown significant effect on chemotherapy-induced myelosuppression.

Difference between GM-CSF and G-CSF

GM-CSF	G-CSF
Stimulates all granulocyte	Stimulates neutrophilis granulocyte
Inhibits neutrophilis granulocyte migration	Enhances neutrophilis granulocyte Migration
Potent activator of macrophages and monocyte	Not an activator of microphages and Monocyte
Induces other cytokines	Does not induce other cytokines

Group of 19 patients who have undergone chemotherapy for breast cancer or melanoma and autologous bone marrow support were treated with GM-CSF. Leucocyte counts obtained 14 days after the transplantation and they were 3120+1744 (32mcg drug/kg/day) and in control group, the count was 863+645/μ1. Treatment with GCSF before chemotherapy of transitional cell carcinoma has shown dose dependent neutrophilis count.

ERYTHROPOIETIN

This drug got approval in 1989 for use in anaemic kidney dialysis patients. Erythropoietin is a glycoprotein hormone produced in kidney and is responsible for the regulations of RBC Production. Erythropoietin production is induced in kidney due to hypoxia, a result of anemia. In chronic renal failure; blood flow to the kidney is decreased continuously resulting in persistent hypoxic state due to incomplete erythropoietic response to anemia associated with chronic renal failure.

Before the advancement of genetic engineering, erythropoietin was obtained from human urine and sheep plasma. The amount is very small in both. The gene for erythropoietin was isolated in 1983 and it was cloned in Chinese hamster ovary cell and allow to express itself to produce erythropoietin. Clinical trails began in 1986 using Epogen and shown a good increase in the number of RBCs in end stage renal disease. The clinical trails proved that epogen was effective, safe, well tolerated drug for treatment of anaemic patients with end stage renal disease. This also reduces the need for blood transfusion and the risk associated with it. Epogen can be used to treat other types of anemias like the one caused in AIDS patients treated with a zidothymidine.

Designing of Drug Delivery Systems

The most existing development in the field of medicine during 1980's has been the development of biotechnology methods to produce newer drugs and also to make known chemotherapeutic agents available in clinically useful quantities. The important techniques of biotechnology include hybridoma technology recombinant DNA technology tissue culture technology and fermentation technology. Using these different techniques, many drugs are produced. Most of the drugs obtained from biotechnology are peptides and polypeptides.

The major challenge to a pharmaceutical scientist is the designing of suitable dosage forms or drug delivery systems for these products. A pharma scientist will come across a number of problems in designing the dosage form. There are three major challenges in this aspect the first challenge is the assurance of purity and potency of these complex molecules. The second challenge is the maintenance of potency and purity of these products as they offer many stability problem. The last challenge is the development of method to improve the absorption of these larger molecules, and prompting their transport and diffusion across the membranes.

The proteins and peptides are highly susceptible to both physical and chemical instability. This poses a lot of problems in purification, storage, formulation and delivery of these products. Chemical instability leads to the formation or cleavage of bonds, thereby forming new chemical entities. Physical instability involves changes in secondary, tertiary or quaternary structure of the molecule. Both Physical and chemical changes can lead to loss of biological activity.

Stability Aspects

Physical Instability

Physical instability in peptides and proteins can be manifested as denaturation, absorption, aggregation and precipitation.

Denaturation

Peptides and polypeptides are made up of both polar and non-polar amino acids. When such peptide come in contact with aqueous environment, the amino acid fold up on themselves to form a globular molecule in which hydrophilic amino acids are exposed to the aqueous medium. If these molecules are then placed in hydrophobic environment (non-aqueous solvent) for purification or formulation, they will unfold, leading to the exposure of hydrophobic amino acids to the environment.

The rearrangements results in loss of tertiary and quaternary structure. This can also occur if solvent is changed from aqueous to mixed solvent is changed from aqueous to mixed solvent (example: alcohol-water, acetone water). Changes in pH also effect the structure of proteins, since this alter the ionisation of carboxylic and amino groups. Similarly, change in ionic strength affect the charge carried by the molecules and lead to denaturation Changes in temperature also cause unfolding of polypeptides and proteins. Increase in temperature breaks the hydrogen bonds and hence, there will be loss of secondary or tertiary structure.

Absorption

Proteins and peptides contain both polar and non-polar amino acids residues (amphillic). Thus, like other surfactants, they also have a tendency to be absorbed at interface, such as air-water, and air-solid. The hydrophobic, non-polar amino acids prefer hydrophobic environment such as air, surfaces of glass or plastic. Thus, peptides and protein may rearrange and denature on absorption at an interface. Once absorbed, they form hydrophobic or van der Waals or ion-pair bonds with the surface.

This may lead to further denaturation. The absorption can occur during storage, processing, formulation, or delivery. Insulin is absorbed to the surface of delivery pumps, to glass and plastic containers and to plastic intravenous bags and tubing. Biological activity is changed or loss due to absorption. This also causes a reduction in the concentration of drug available.

Aggregation and precipitation

Denatured, unfolded protein may rearrange to form

aggregates in which hydrophobic amino acid reduces of different molecules are associated together. More amount of aggregation leads to precipitation. Interfacial absorption leads to precipitation and aggregation. The process is enhanced by the presence of hydrophobic surfaces. Presence of large air-water interface can accelerate precipitation. Agitation of protein and polypeptides can also lead to aggregation. This is due to introduction of air-bubbles as well as increased thermal motion of the molecules.

Chemical Instability

There is different reaction of proteins and polypeptides, which leads to chemical instability. There are listed below.

Factors responsible for chemical responsibility in proteins and polypeptides

Hydrolysis	Oxidation
Imide formation	Isomerisation
Deamidation	Disulphide Exchange
Racemisation	Photodecomposition

Deamidation

It involves hydrolysis of the side chain of amide linkage of an amino acid reduce to form a carboxylic acid. Some common proteins, which undergo in vitro deamidation, are human growth hormone insulin and prolactin. Glutamine and asparagines are the amino acid, which undergo deamidation. The rate of deamidation is increased by an increase in pH, temperature and ionic strength. The tertiary strength of protein resists deamidation, For example, tertiary structure of trypsin prevents deamidation. Deamidation reduces biological activity. For example, the activity of ACTH (adrenal corticotrophin hormone) is reduced by deamidation.

Oxidation

Oxidation occurs in the side chains of histidine, methionine, lysine, tyrosine and tryptophan resides in protein. It is commonly during synthesis, isolation and storage. Atmospheric oxygen oxidises methionine under acidic conditions. Oxidising agents like hydrogen peroxide, iodine and dimethyl sulphoxide are also responsible for oxidation. Oxidations are reduced biological activity. For example, oxidation of methionine in gastrin, and corticotrophin results in loss of activity. In some proteins such as Lysozyme, biological activity can be restored by reduction. Some other proteins such as glucagons can retain the activity even after oxidation.

Racemisation

Except glycine, all other amino acids are chiral at carbon bearing the side chain and hence are susceptible to Racemisation. Thus racemisation may convert the protein non-metabolisable because the racemic peptide bonds are inaccessible to proteolytic enzymes and can reduce biological activity.

Disulphide exchange

Breaking and incorrect reformation of disulphide bonds may alter the three-dimensional structure of a protein and hence can change or alter biological activity. The reaction is catalysed by thiols, which arise as a result of hydrolytic cleavage of disulphides.

Proteolysis

Proteolysis is the cleavage of protein molecule by the breakage of peptide bond. This depends on the residues involved. Asparagine residue, particularly the bond between asparagines and proline is highly susceptible to cleavage. But the peptide bonds, in general, are stable at neutral pH and room temperature. Proteolysis occurs upon heating. For example; heating at 90-100°C inactivates lysozyme.

Stability Improvements in Polypeptides

Physical stability

Certain additives given in table below can be used to improve physical stability of polypeptides and proteins. Specific ion-binding sites introduced on the protein molecules leads to improve stability. Both nonionic and iconic surfactants can stabilise proteins by preventing their absorption at the interface. They also inhibit aggregation and precipitation Salts can decrease denaturation and increase the stability by nonspecific or specific ion binding to the protein.

Methods used to improve physical and chemical stability of peptides and proteins

Physical stability	Chemical stability
Additives used	Chemical modification of molecules
Polyalcohols	Site directed mutagenesis
Salts	Appropriate choice of conditions
Surfactants	
Metal ions	
Chelating agents	
Reducing agents	
Fatty acids	
Phospholipids	
Disulphide exchange	

At low concentration in aqueous solution, polyhydric alcohol such as glycerol can stabilise proteins by selective salvation. Here water molecules pack around the protein to exclude the hydrophobic alcohol molecules and increase stability. The reverse occur at high concentration of the polyalcohol.

Chemical stability

Different methods are used to improve chemical stability. They are described below.

Modification of Protein Molecules: Synthetic polymers like PEG and polyoxyethylene can be coupled to protein molecules chemically to enhanced stability. However, there are chances of reduction or complete loss of biological activity. Comparatively, lipids are better coupling agents. Lipid coupled with insulin can increase the absorption of insulin and hence increase the activity.

Site Directed Mutagenesis: These techniques created specific mutations that alter the amino acids sequence of proteins and creates new proteins. The process can be designed to improve both stability as well as specificity of proteins. Here, specific amino acids, which are reactive, can be replaced with others, which are non-reactive, while retaining the biological activity of the protein. It is also possible to introduce specific nonbinding sites to improve the stability. Disulphide bonds can be introduced, which can stabilise the original confirmation and provide stability against denaturation.

Choice of Conditions: Chemical stability can be obtained by appropriate choice of conditions, such as physical state, pH temperature, preservatives and ionic strength Peptides and proteins can be stabilised if they are in dry state. Lower temperature also enhanced stability. Dosage forms can be formulated in a way that they are solid and freezedried products. At the time of administration, they can be reconstituted. The stability of proteins and polypeptides is an important consideration in the selection of route of administration and appropriate dosage form.

Routes of Administration

Oral route

Oral route is the most convenient route for administration of drugs. But this route is problematic for protein and peptides drugs Peptides and proteins will be degraded by enzymatic action in the GIT. These molecules cannot cross the epithelial membrane due to their high molecular weight. If they are absorbed they undergo first pass metabolism in the liver Proteins and polypeptides undergo hydrolysis GIT to form a mixture of smaller peptides and amino acids. But small peptides such as dipeptides and tripeptides remain stable in GIT (example: TRH thyrotropin releasing hormone).

Although these small peptides and a few larger peptides such as cyclosporin (undeca peptide) can remain stable, they are not at all observed to a significant extent. The bioavailability of TRH is only a few percent. The gastric fluids, is intestinal fluids and the digestive juices contain many proteolytic enzymes. These enzymes hydrolyte the larger peptides to absorbable product. But as stated earlier, a few smaller peptides such As TRH and cyclosporin are absorbed intact. This absorption is done through carrier-mediated transport. Most of the drugs are absorbed by passive transport. Here the rate of absorption is a function of many factors like pKa of the drug, particle size, and partition coefficient, stability in GIT, dose to solubility ratio, and molecular weight of the drug.

High molecular weight substances such as peptides, which have molecular weight above 500, are very poorly absorbed in GIT. Also the mucin layer acts as a physical barrier for the absorption these molecules can occur by the following mechanisms. Water-soluble molecules can diffuse through intercellular spaces (or water pores). Lipid soluble molecules can diffuse through the membrane lipids. A Water-soluble as well as lipid soluble materials can enter into epithelial cells by pinocytosis or phagocytosis. But the absorption by these mechanisms is very less, approximately 1-5 percent. Some peptides, which are absorbed by facilitated transport, are absorbed in greater extent. But this is also insufficient to elicit a pharmacological action.

Oral Bioavailability of Proteins and Peptides

The oral bioavailability of peptides and protein can be improved by the following methods:

Peptidase inhibitors

Protection against the hydrolysis can be given by co-administration of peptidase inhibitors with proteins and peptides. The enhanced the stability of peptides for example, amistatin acts as a peptidase inhibitor and promotes the absorption of enkephalins

Saturation of peptidases

The saturation of peptidases can protect the peptides. This can be achieved by either increasing the dose or by administrating another protein or peptide with the drug. Increasing the dose can do saturation of peptidases. But this is possible only if the drug is non-toxic at high concentration. The drugs may be formulated in an enteric coated form to bypass some of the peptidases in GIT. Polymeric coating cross linked with azoaromatic groups from a tough water-resistant film, which remains intact in stomach and small intestine. The natural microbial flora of colon reduces the azo bonds and hence the drug will be released in large intestine. But the colon is much less permeable to water-soluble drugs then small intestine and this reduces the bioavailability of peptides.

Penetration enhancers

By the co-administration of penetration enhancers, the absorption of proteins and peptides can be improved. Different penetrations enhancers include surfactants, non-surfactants and chelating agents. Surfactants such as bile salt and detergents increase the membrane fluidity and hence increase the absorption of proteins and peptides. Micelles formed by the surfactants may interact with the membrane phospholipids of intestine and promote the absorption. Also, pericellular permeation is increased by solubilisation of drugs in micelles formed by bile salts. Bile salt micelles are reported to enhance the absorption of insulin significantly.

Chelating agents such as EDTA and sodium citrate have been proved to enhance the absorption of insulin and heparin. They are supposed to promote the absorption of water soluble drug by chelating the calcium and magnesium ions of epithelial tissue and hence, cause their depletion in the regions of tight junction between the epithelial cells. This will force the water by osmosis and enhances pericellular absorption of peptides and proteins.

Lymphatic Uptake

A part from enzymatic degradation and poor absorption of peptides from GIT, there is another important factor, that is the first pass metabolism in the liver. The hepatitis first pass metabolism can be avoided by two ways:

— By blocking specific enzymes

— By promoting lymphatic uptake

The lymphatic uptake may be enhanced by surfactants, and bile salts, which facilitate micellar solubilisation. This can also be achieved by using lipid vehicles or by simultaneous feeding of lipid. The lipid enhances absorption via lymph system. The lipid designation lead to the formation of chylomicrons, in which the drug may be the trapped and selectively absorbed by the lymphatic system Salicylates are also known to enhance lymphatic uptake of peptides. The more hydrophilic is the salicylate, the greater is the activity.

Parental Route

Parental route is the practical method for administration of proteins. Here also the drug can be targeted to a particular tissue in a controlled release manner. Parental targeting is done for the following reasons. Site specific targeting will improve the therapeutic index of the drug by optimising the access, and the nature of interaction with receptors. Targeting protects the drugs as well the body from unwanted and deleterious disposition. The rate of release can also be controlled and this allows the drugs with short biological half-life (proteins and peptides) and drugs with narrow therapeutic indices to be utilised. A parental controlled release system maintains a constant drug level in the blood and eliminates the problems of patient's compliance.

Genetically engineered proteins and peptides are ideal candidates for targeting. These have a very short biological half-life in the blood. For example, angiotensin has a half-life of 15 seconds and oxytocin has half-life of 2 minutes. Similarly the half-life of vasopressin is only 4 minutes and that of insulin is less than 25 minutes. The peptide and protein drugs are chemically similar to the endogenous substances they are intended to replace. Hence their metabolism in the body becomes easy. Both plasma and tissue peptides can breakdown these drugs.

If they are to undergo enzymatic degradation in plasma, they may cause immunological reactions and side effect. Thus by encapsulating the proteins and peptides in a target able drug delivery system, they can be protected from enzymatic degradation until they reach the target tissue or site of action. The biotechnology products are highly potent If they are given in convention dosage forms, whole body will be exposed to them. Also due to distribution through out the body, large quantities are required to achieve therapeutic level. The proteins and peptides, especially at high concentration can stimulate immune system. These molecules cannot pass through various membranes of our body due to large molecular weight formulating them in controlled release target able dosage forms will solve all these problems.

Parental Targeting

Parental drug carrier systems are of two types namely, particulate carriers and soluble carriers. Particle size and surface properties control the fate of colloidal particles administrated intravenously. Particle larger then 7 μm are entrapped in fine capillaries of lungs. Particles of size range between 100nm and 7μm, which are hydrophobic and carry a negative charge, are taken up by reticuloendothelial system (RES), if they are foreign. Such uptake of particles by the lungs and RES provides opportunity for passive targeting.

Else attaching a site-specific moiety to the carrier can actively target the blood. Passive targeting is the simplest method of parental targeting. Passive targeting to RES can be utilised to treat RES diseases such as viral and bacterial infection and neoplasm's. Microphage activating factor for the activation of macrophages to kill the tumors can be formulated into liposome's, which will be taken up by the macrophages. Muramyl dipeptide is currently under study in a liposomeencapsulated form as a macrophage-activating factor.

Inactive targeting, either the drug or the carrier is coupled to a site-specific moiety such as native immunoglubolin, erythrocyte membrane glycoprotein or MAbs. Tumor reactive MAbs are used to target cytotoxic drugs to neoplastic tumors; Synthetic polymers can be chemically conjugated to peptides and proteins to alter pharmacodynamics. They can increase protein size or reduce unwanted interaction with blood and tissue components. The conjugation with hydrophilic polymers reduces immunogenecity and increase duration of action. Many problems are associated with parental targeting. Some of them are given below.

1. If targeting is done with particulate carrier, sequestration by RES is the main problem and due to this, the drugs may not reach the site of action.
2. The particles cannot cross the vascular endothelium. Very small particles and some large molecules can pass through. Hence the particulate system should release the drug outside the cell. This leads to metabolism, prior to uptake by the cells.
3. The transport of drugs can occur by either pinocytosis or phagocytosis.

Receptor mediated transcytotic routes exist for only a few macromolecules. Thus, intercellular transport of macromolecules through cell membranes is a major obstacle.

Parental Targeted Drug Delivery Systems

The parental targeted drug delivery systems for proteins and peptides are either soluble carrier systems or particulate carrier systems.

Soluble carrier systems

Complexation of peptides and proteins with polymers and

lipids enhances the stability and site specificity, while retaining biological activity. Specific properties of target cell are exploited for targeting. For example, MAbs or other antibodies specific for a variant cell surface antigenic determinant can be complexed with drugs. The drug conjugate may be active at the cell surface, or may become active after endocytosis, or the conjugate may release the drug extracellularly in a controlled manner.

Genetic engineering can be utilised to have site specific sophisticated protein molecule. Hybrid fusion technique has also been used. Such molecule (Hybrid protein) combine future of two to three proteins form drug with target recognition properties as well as pharmacological activity. They have increased half-life, altered immunogenetically, and increased stability.

Particulate carrier systems

Various particulate systems, which can be used as carriers for targeting of peptides and proteins. These can be targeted passively (according to their particle size) or actively (by conjugating with a tractable moiety like MAb).

Liposomes: these are tiny spheres formed when phospholipids are combined with water. The structure of liposomes resembles that of cell membranes. They are vesicles made up of either one or more lipid bilayer alternating with aqueous compartments. Liposome's are of the following types:

— Small unilamellar vesicles (SUV) they have a single lipid bilayer of the size of 20.70nm.

— Large unilamellar vesicles (LUV) size 100-400nm.

— Multilamellar vesicles (MLV) 200nm to several microns size and contain two or more concentric bilayer.

In liposome's, various molecules can be incorporated to modify their properties. For example, incorporation of cholesterol into the liposome's result in more stable membrane. Liposome's can be passively targeted to liver. SUVs can be targeted to In liposome's, various molecules can be incorporated to modify their properties. For example, incorporation of cholesterol into the liposome's result in more stable membrane.

Liposome's can be passively targeted to liver. SUVs can be targeted to parenchymal cells of liver due to their small size. They can also be targeted to other organs by attaching immunoglobulins to them. They are being tested for controlled release of genetically engineered peptides and protein such as macrophage stimulating agent, muramyl dipeptide and enzymes. The use of liposome's for relatively less stable, the entrapped drug may leak and their drug carrying capacity is low.

Microspheres: They are solid spherical particles with a size range of 1 to 600μm and contain dispersed drug. They are prepared by emulsification; here the dispersed phase consists of droplets of polymer-drug solution. Otherwise, the polymer droplet can be cross-linked chemically, thermally or by gamma radiation. Entrapment or ionic or covalent binding can be prepared either by natural biodegradable polymers (Gelatin or albumin) or by synthetic polymers (polylactic or polyglycolic acid). Two types of microspheres are possible in first type, the drug is encapsulated in a microcapsule and in the second type, the drug is dispersed in polymer matrix. The Microspheres can be passively targeted to lung capillaries or to RES. Active targeting is also possible if they are attached to targeting moieties. Albumin Microspheres mixed with iron oxide can be used as magnetic Microspheres and can be targeted to specific organs by the use of an external magnet.

Nanoparticles: These are similar to Microspheres but have a particle size ranging between 10 and 1000nm. These can be made from biodegradable material such as albumin such as methyl methacrylate. Targeting the Nanoparticles with targeting the moieties such as monoclonal antibodies increases their specificity of targeting.

Released Erythrocytes: When red blood cells are suspended in hypotonic solution permeability increases. If the solutions contains dissolved peptides and proteins, they form equilibrium on either side of the membrane. Then changing the ionic concentration of solution to isotonicity and incubating at 37°c restabilises the membrane. The pores will close and make the erythrocytes to reseal. Using this technique, up to 40 percent of drug can be entrapped in RBCs. The released RBCs can be targeted to liver or spleen because aged, damaged or modified RBCs are taken up by RES and sent to liver. These can be used for enzymes replacement in genetic metabolic disorder. To be targeted by released RBCs. The drug should not be denatured under hypotonic conditions.

Microemulsions: Colloid sized emulsion droplets can be used for targeting as well as for controlled release of peptides and proteins. The stability and fate of parental emulsion depends upon the emulsifier used. Most commonly used emulsifying agents in lecithin. RES takes up the emulsion droplets. Vegetable oil emulsions are preferentially deposited in adipose tissue, mammary glands and myocardial muscle and hence can be used to target the drug to these tissues.

Transdermal Route

Till recent times, skin was used only for the treatment of local infections. However, a better understanding of skin structure and functions has led to the utilisation of this tissue for systemic delivery of drugs. The transdermal route has many advantages. They include:

— Avoidance of first pass metabolism,

— Reduction in frequency of dosing,

— Easy termination of dosing

— Increased patient compliance

The common disadnantages of this route are:

1. Very low rate of permeation for most of the drugs
2. High variability in intra and inter subject permeation

Peptides and proteins have the difficult of penetration through stratum corneum due to their molecular weight; ionic nature and hydrophilic nature. The penetration can be enhanced by iontophoresis, and utilisation of permeation enhancers.

Iontophoresis

It is the migration of ions when an electric current is passed through a solutions containing ionised species. Drugs in ionic form can be "phoresed" into the body using small current below the pain threshold of the patients. The drug is placed in a gauze pad and is applied on the skin. An electrode is placed at the distant part of the body and the current is allowed to flow for sufficient time period. To undergo iontophoresis, the proteins and peptides should carry charge. Charge can be included on the protein molecule by controlling the pH and ionic strength of the solution.

Insulin has been successfully studied for iontophoresis through intact rate skin. The rate and extent of transdermal absorption of TRH has been improved by iontophoresis are burning of skin and denaturation of peptides and proteins. They are denatured during charging. Due to current and also due to the heat generated during iontophoresis.

Phonophoresis

In this method, ultrasonic sound is applied to the coupling contact agent. The ultrasonic waves increase the drug absorption by thermal effects and also by temporary alteration of the physical structure of the skin. In this method also the protein may get denatured due to the heat generation.

Penetration Enhancers

The substances such as a zone and oleic acid enhance the drug absorption through the skin. They fluidise the intercellular lipid lamellae of stratum corneum. Fatty acids disrupt the packed structure of lipids in the extra cellular spaces. The major problem associated with the use of penetration enhancers is the skin irritation caused by them.

Prodrugs

Several proteins and peptides including TRH, neurotensin and fibrinopeptides posses a terminal pyroglutamyl residue This can be easily derived to produce prodrugs with enhanced permeability across the skin. The skin contains enzymes, which hydrolyse the prodrug at physiological pH to generate the active drug.

Nasal Route

The nasal route is another option for delivery of peptides and proteins. This route of administration has many advantages, which are listed below:

— It is convenient for administration

— Self-medication is possible

— The surface area is large and mucosa is highly vascularised

— This route bypasses the first pass metabolism

When compared to other routes, this route shows high permeability for peptides and proteins. The bioavability of peptides through nasal route depends on molecular weight and physicochemical properties. It is of the order of 1-20 percent. Some of the disadvantages of nasal route include the following:

— Variation in the mucous secretion effects the extent of absorption

— Peptidases present in nasal membrane degrade the peptides.

— The absorption will be altered with diseases state such as upper respiratory tract infections, allergic rhinitis and common cold.

— Penetration enhancers used may caused damage to mucosal cell membrane and may be ciliotoxic.

The clearance of drugs from nasal cavity is very fast in posterior nasal cavity. Here it is of the order of 10-15 minutes. Thus, nasal sprays are designed to deliver the drugs to the anterior cavity to reduce the clearance. The absorption of proteins and peptides through nasal mucosa can be enhanced by penetration enhancers. They include surface active agents. These penetration enhancers increase the fluidity of lipid bilayer of the membrane, and also open the aqueous pores as a result of calcium chelation, and thus increase the penetration.

Peptides inhibitors also increase the absorption by decreasing the peptidase activity. But they cause damage to the mucosal membrane and are ciliotoxic in nature. Some penetration enhancers like oleic acid, salicylic acid, linoleic acid and sodium taurodihydrofusidate are safe. Damage to nasal epithelium, which is defence layer against infectious organisms, may lead to severe problems. Thus the penetration enhancer should leave the nasal defence system intact. The nasal route is successfully used for administration of enkaphalin analogous, insulin and interferon's. It can also be used for administration of pituitary hormones such as vasopressin and oxytocin. However, this route is suitable for long-term use and is short term delivery of peptides and proteins.

Rectal Route

The large intestine receives the drainage mainly by hepatic portal vein. The lower portion receives drainage by lymphatic. This route is particularly suitable for peptides and proteins because these show greater lymphatic uptake due to their high molecular weight. The first pass metabolism by liver is also avoided. Also the rectal route avoids most of the GIT peptidases.

The bioavailability of peptides and protein from this route is particularly low because of the tight intercellular junctions of the rectal epithelium. Hence, penetration enhancers are required. Surfactants, if used as penetration enhancers cause bleeding and mucosal damage If silicates are used, the above problem is solved. Designing of suitable suppository can be done for peptides and proteins along with a mild penetration enhancer.

Buccal Route

This route has many advantages for the delivery of peptides and proteins.

— The sensitivity is less when compared to nasal and other routes.

— The drug delivery system can be removed at any time. This route has better patient compliance.

But the absorption of proteins and peptides is less when compared to nasal route. The common method for delivery is the tablet form. But there are chances of drug loss due to swallowing of saliva or designing a self-adhesive dosage from, which can retain in the oral cavity in contact with buccal mucosa, can solve accidental swallowing of tablet. These problems like water-soluble and water insoluble, ionic as well as non-ionic polymers can be used for buccal adhesion. Polyacrylate hydro gels are most commonly used, the adhesive polymer can act as drug carrier. In a second type of device, it acts as adhesive link between disk loaded with drug and the mucosa.

In a third party of device, the disk is fixed to mucosa by self adhesive polymer shield. This device can be applied for several days. But they interface with eating, drinking as well as talking. So, the maximum time of application should be a few hours. Though peptides and proteins from biotechnology can be delivered by all the above-mentioned routes, the main limitation is the requirement of penetration enhancer. Milder, less irritating and less damaging penetration enhancer should be developed for the non-conventional routes of administration such as transdermal, nasal, rectal and buccal routes. Among the non-conventional routes, buccal route appears to be more promising because of its least susceptibility to irritation.

Future of Biotechnology

— The future of biotechnology promises many advances in drugs, drug delivery systems, diagnostic kits, and other pharmaceutical specialties, including polypeptides, proteins and antibiotics.

— It also provides a basis for understanding the mechanisms & processes involved in diseases like causes of cancer, neurological, neuromuscular and cardiac diseases can be better understood. This understanding helps in better treatment and prevention of diseases.

— Biotechnology provides new avenues for improvement of quality of human life. Thus, biotechnology is helping the world to move towards a future, which is safer & healthier.

— The process of drug synthesis has been made easy by biotechnology.

— Many of the reactions can be carried out biologically by the use of plant or animal tissue cultures and microorganisms.

10

Biotechnological Applications in Environment

Our lives may, quite literally, be changed. Environmental biotechnology deals with far less apparently dramatic topics and, though their importance, albeit different, may be every bit as great, their direct relevance is far less readily appreciated by the bulk of the population. Cleaning up contamination and dealing rationally with wastes is, of course, in everybody's best interests, but for most people, this is simply addressing a problem which they would rather had not existed in the first place.

Even for industry, though the benefits may be noticeable on the balance sheet, the likes of effluent treatment or pollution control are more of an inevitable obligation than a primary goal in themselves. In general, such activities are typically funded on a distinctly limited budget and have traditionally been viewed as a necessary inconvenience. This is in no way intended to be disparaging to industry; it simply represents commercial reality. In many respects, there is a logical fit between this thinking and the aims of environmental biotechnology.

For all the media circus surrounding the grand questions of our age, it is easy to forget that not all forms of biotechnology involve xenotransplantation, genetic modification, the use of stem cells or cloning. Some of the potentially most beneficial uses of biological engineering, and which may touch the lives of the majority of people, however indirectly, involve much simpler approaches. Less radical and showy, certainly, but powerful tools, just the same.

Environmental biotechnology is fundamentally rooted in waste, in its various guises, typically being concerned with the remediation of contamination caused by previous use, the impact reduction of current activity or the control of pollution. Thus, the principal aims of this field are the manufacture of products in environmentally harmonious ways, which allow for the minimisation of harmful solids, liquids or gaseous outputs or the clean-up of the residual effects of earlier human occupation. The means by which this may be achieved are essentially two-fold.

Environmental biotechnologists may enhance or optimise conditions for existing biological systems to make their activities happen faster or more efficiently, or they resort to some form of alteration to bring about the desired outcome. The variety of organisms which may play a part in environmental applications of biotechnology is huge, ranging from microbes through to trees and all are utilised on one of the same three fundamental bases - accept, acclimatise or alter. For the vast majority of cases, it is the former approach, accepting and making use of existing species in their natural, unmodified form, which predominates.

Accordingly, the range of businesses to which environmental biotechnology has potential relevance is almost limitless. One area where this is most apparent is with regard to waste. All commercial operations generate waste of one form or another and for many, a proportion of what is produced is biodegradable. With disposal costs rising steadily across the world, dealing with refuse constitutes an increasingly high contribution to overheads. Thus, there is a clear incentive for all businesses to identify potentially cost-cutting approaches to waste and employ them where possible.

Changes in legislation throughout Europe, the US and elsewhere, have combined to drive these issues higher up the political agenda and biological methods of waste treatment have gained far greater acceptance as a result. For those industries with particularly high biowaste production, the various available treatment biotechnologies can offer considerable savings. Manufacturing industries can benefit from the applications of whole organisms or isolated biocomponents. Compared with conventional chemical processes, microbes and enzymes typically function at lower temperatures and pressures.

The lower energy demands this makes leads to reduced costs, but also has clear benefits in terms of both the environment and workplace safety. Additionally, biotechnology can be of further commercial significance by converting low-cost organic feedstocks into high value products or, since enzymatic reactions are more highly specific than their chemical counterparts, by deriving final substances of high relative purity. Almost inevitably, manufacturing companies produce wastewaters or effluents, many of which contain biodegradable contaminants, in varying degrees.

Though traditional permitted discharges to sewer or watercourses may be adequate for some, other industries, particularly those with recalcitrant or highly concentrated effluents, have found significant benefits to be gained from using biological treatment methods themselves on site. Though careful monitoring and process control are essential, biotechnology stands as a particularly cost-effective means of reducing the pollution potential of wastewater, leading to enhanced public relations, compliance with environmental legislation and quantifiable cost-savings to the business.

Those involved in processing organic matter, for example, or with drying, printing, painting or coating processes, may give rise to the release of volatile organic compounds (VOCs) or odours, both of which represent environmental nuisances, though the former is more damaging than the latter. For many, it is not possible to avoid producing these emissions altogether, which leaves treating them to remove the offending contaminants the only practical solution. Especially for relatively low concentrations of readily water-soluble VOCs or odorous chemicals, biological technologies can offer an economic and effective alternative to conventional methods. The use of biological cleaning agents is another area of potential benefit, especially where there is a need to remove oils and fats from process equipment, work surfaces or drains.

Aside from typically reducing energy costs, this may also obviate the need for toxic or dangerous chemical agents. The pharmaceutical and brewing industries, for example, both have a long history of employing enzyme-based cleaners to remove organic residues from their process equipment. In addition, the development of effective biosensors, powerful tools which rely on biochemical reactions to detect specific substances, has brought benefits to a wide range of sectors, including the manufacturing, engineering, chemical, water, food and beverage industries.

With their ability to detect even small amounts of their particular target chemicals, quickly, easily and accurately, they have been enthusiastically adopted for a variety of process monitoring applications, particularly in respect of pollution assessment and control. Contaminated land is a growing concern for the construction industry, as it seeks to balance the need for more houses and offices with wider social and environmental goals. The reuse of former industrial sites, many of which occupy prime locations, may typically have associated planning conditions attached which demand that the land be cleaned up as part of the development process.

With urban regeneration and the reclamation of 'brown-field' sites increasingly favoured in many countries over the use of virgin land, remediation has come to play a significant role and the industry has an ongoing interest in identifying cost-effective methods of achieving it. Historically, much of this has involved simply digging up the contaminated soil and removing it to landfill elsewhere. Bioremediation technologies provide a competitive and sustainable alternative and in many cases, the lower disturbance allows the overall scheme to make faster progress.

The UK's Department of Trade and Industry estimated that 15-20% of the global environmental market in 2001 was biotech-based, which amounted to about 250-300 billion US dollars and the industry is projected to grow by as much as ten-fold over the following five years. This expected growth is due to greater acceptance of biotechnology for clean manufacturing applications and energy production, together with increased landfill charges and legislative changes in waste management which also alter the UK financial base favourably with respect to bioremediation.

Biotechnology-based methods are seen as essential to help meet European Union (EU) targets for biowaste diversion from land-fill and reductions in pollutants.

Across the world the existing regulations on environmental pollution are predicted to be more rigorously enforced, with more stringent compliance standards implemented. All of this is expected to stimulate the sales of biotechnology-based environmental processing methods significantly and, in particular, the global market share is projected to grow faster than the general biotech sector trend, in part due to the anticipated large-scale EU aid for environmental clean-up in the new accession countries of Eastern Europe. Other sources paint a broadly similar picture.

The BioIndustry Association (BIA) survey, Industrial Markets for UK Biotechnology - Trends and Issues, published in 1999 does not quote any monetary sector values per year, but gives the size of the UK sector as employing 40 000 people in 1998 with an average yearly growth over 1995-98 of 20%. Environmental biotech is reported as representing around 10% of this sector.

The Organisation for Economic Cooperation and Development (OECD) estimates that the global market for environmental biotechnology products and services alone will rise to some US$75 billion by the year 2000, accounting for some 15 to 25% of the overall environmental technology market, which has a growth rate estimated at 5.5% per annum. The UK potential market for environmental biotechnology products and services is estimated at between 1.65 and 2.75 billion US dollars and the growth of the sector stands at 25% per annum.

The Organisation for Economic Cooperation and Development concluded that the industrial use of biotechnology commonly leads to increasingly environmentally harmonious processes and additionally results in lowered operating and/or capital costs. For years, industry has appeared locked into a seemingly unbreakable cycle of growth achieved at the cost of environmental damage. The OECD investigation provides what is probably the first hard evidence to support the reality of biotechnology's long-heralded promise of alternative production methods, which are ecologically sound and economically efficient.

A variety of industrial sectors including pharmaceuticals, chemicals, textiles, food and energy were examined, with a particular emphasis on biomass renewable resources, enzymes and bio-catalysis. While such approaches may have to be used in tandem with other processes for maximum effectiveness, it seems that their use invariably leads to reduction in operating or capital costs, or both. Moreover, the research also concludes that it is clearly in the interests of governments of the developed and developing worlds alike to promote the use of biotechnology for the substantial reductions in resource and energy consumption, emissions, pollution and waste production it offers.

The potential contribution to be made by the appropriate use of biotechnology to environmental and economic sustainability would seem to be clear. The upshot of this is that few biotech companies in the environmental sector perceive problems for their own business development models, principally as a result of the wide range of businesses for which their services are applicable, the relatively low market penetration to date and the large potential for growth. Competition within the sector is not seen as a major issue either, since the field is still largely open and unsaturated.

Moreover, there has been a discernible tendency in recent years towards niche specificity, with companies operating in more specialised subarenas within the environmental biotechnology umbrella. Given the number and diversity of such possible slots, coupled with the fact that new opportunities, and the technologies to capitalise on them, are developing apace, this trend seems likely to continue. It is not without some irony that companies basing their commercial activities on biological organisms should themselves come to behave in such a Darwinian fashion. However, the picture is not entirely rosy.

Typically the sector comprises a number of relatively small, specialist companies and the market is, as a consequence, inevitably fragmented. Often the complexities of individual projects make the application of 'standard' off-the shelf approaches very difficult, the upshot being that much of what is done must be significantly customised. While this, of course, is a strength and of great potential environmental benefit, it also has hard commercial implications which must be taken into account. A sizeable proportion of companies active in this sphere, have no products or services which might reasonably be termed suitable for generalised use, though they may have enough expertise, experience or sufficiently perfected techniques to deal with a large number of possible scenarios.

The fact remains that one of the major barriers to the wider uptake of biological approaches is the high perceived cost of these applications. Part of the reason for this lies in historical experience. For many years, the solutions to all environmental problems were seen as expensive and for many, particularly those unfamiliar with the multiplicity of varied technologies available, this has remained the prevalent view. Generally, there is often a lack of financial resource allocation available for this kind of work and biotech providers have sometimes come under pressure to reduce the prices for their services as a result.

Greater awareness of the benefits of biotechnology, both as a means to boost existing markets and for the opening up of new ones, is an important area to be addressed. Many providers, particularly in the UK, have cited a lack of marketing expertise as one of the principal barriers to their exploitation of novel opportunities. In addition, a lack of technical understanding of biotech approaches amongst target industries and, in some cases, downright scepticism regarding their efficacy, can also prove problematic. Good education, in the widest sense, of customers and potential users of biological solutions will be one major factor in any future upswing in the acceptance and utilisation of these technologies.

Effect of Local Circumstances

Another of the key factors affecting the practical uptake of environmental biotechnology is the effect of local circumstances. Contextual sensitivity is almost certainly the single most important factor in technology selection and represents a major influence on the likely penetration of biotech processes into the marketplace. Neither the nature of the biological system, nor of the application method itself, play anything like so relevant a role. This may seem somewhat unexpected at first sight, but the reasons for it are obvious on further inspection. While the character of both the specific organisms and the engineering remain essentially the same irrespective of location, external modalities of economics, legislation and custom vary on exactly this basis.

Accordingly, what may make abundant sense as a biotech intervention in one region or country, may be totally unsuited to use in another. In as much as it is impossible to discount the wider global economic aspects in the discussion, disassociating political, fiscal and social conditions equally cannot be done, as the following example illustrates. In 1994, the expense of bioremediating contaminated soil in the United Kingdom greatly exceeded the cost of removing it to landfill. Six years later, with successive changes of legislation and the imposition of a landfill tax, the situation has almost completely reversed.

In those other countries where landfill has always been an expensive option, remediation has been embraced far more readily. While environmental biotechnology must, inevitably, be viewed as contextually dependent, as the previous example shows, contexts can change. In the final analysis, it is often fiscal instruments, rather than the technologies, which provide the driving force and sometimes seemingly minor modifications in apparently unrelated sectors can have major ramifications for the application of biotechnology.

Increasingly tough environmental law makes a significant contribution to the sector and changes in regulatory legislation are often enormously influential in boosting existing markets or creating new ones. When legislation and economic pressure combine, as, for example, they have begun to do in the European Landfill Directive, the impetus towards a fundamental paradigm shift becomes overwhelming and the implications for relevant biological applications can be immense. There is a natural tendency to delineate, seeking to characterise technologies into particular categories or divisions.

However, the essence of environmental biotechnology is such that there are many more similarities than differences. Though it is, of course, often helpful to view individual technology uses as distinct, particularly when considering treatment options for a given environmental problem, there are inevitably recurrent themes which feature throughout the whole topic. Moreover, this is a truly applied science. While the importance of the laboratory bench cannot be denied, the controlled world of research translates imperfectly into the harsh realities of commercial implementation. Thus, there can often be a dichotomy between theory and application and it is precisely this fertile ground which is explored in the present work.

In addition, the principal underlying approach of specifically environmental biotechnology, as distinct from other kinds, is the reliance on existing natural cycles, often directly and in an entirely unmodified form. Thus, this science stands on a foundation of fundamental biology and biochemistry. To understand the application, the biotechnologist must simply examine the essential elements of life, living systems and ecological circulation sequences. However engineered the approach, this fact remains true. In many respects, environmental biotechnology stands as the purest example of the newly emergent bioindustry, since it is the least refined, at least in terms of the basis of its action.

In essence, all of its applications simply encourage the natural propensity of the organisms involved, while seeking to enhance or accelerate their action. Hence, optimisation, rather than modification, is the typical route by which the particular desired end result, whatever it may be, is achieved and, consequently, a number of issues feature as common threads within the discussions of individual technologies.

INTEGRATION

Integration is an important aspect for environmental biotechnology. The wider goal of integration is not, of necessity, confined solely to the specific methods used. In some spheres, traditional biology has become rather unfashionable and the emphasis has shifted to more exciting sounding aspects of life science.

While the new-found concentration on 'ecological processes', or whatever, sounds distinctly more 'environmental', in many ways, and somewhat paradoxically, it sometimes serves the needs of environmental biotechnology rather less well. The fundamentals of living systems are the stuff of this branch of science and, complex though the whole picture may be, at its simplest the environmental biotechnologist is principally concerned with a relatively small number of basic cycles. In this respect, a good working knowledge of biological processes like respiration, fermentation and photosynthesis, a grasp of the major cycles by which carbon, nitrogen and water are recycled and an appreciation of the flow of energy through the biosphere must be viewed as prerequisites.

Forest Biotechnology

Environmental issues are intertwined with the whole fabric of society and have many facets-social, political, economic, scientific, and ethical. Biologically based processes can be used in the pulp and paper industry to reduce some negative environmental impacts. Similarly, forest biotechnology can solve some problems faced by forest managers. Effective management of forested lands is central to our quality of life and the sustainability and health of the planet. Policies, or a lack of policies, in one part of the world cannot be isolated from their impact on the global community. We need to be concerned with tree improvement as it relates to forest health, biodiversity, sustainability, resiliency, and other conditions linked to the global forest resource.

While new breeding techniques, fertilisers, pesticides, and improved cultural methods are conventional ways to improve productivity, genetic engineering is a more controversial alternative. However, biotechnology has the potential for generating forest tree cultivars that cannot be produced by conventional breeding alone. Biotechnological approaches are being investigated for integrating conventional forest tree breeding with forest resource productivity.

Biotechnology in Paper Industry

Biotechnology has provided environmentally sound solutions to problems faced by the papermaking industry. Kuhad et al. attribute the importance of biotechnology to its potential for more specific reactions, less environmentally deleterious processes, energy savings, and capacity to be used in place of nonbiological processes. Biological products have long played a critical role in paper mill sludge treatment and effluent clarification. Newer biological applications include targeting organic contaminants in thermomechanical pulp (TMP) mill white water for fungal treatment and dechlorinating kraft bleach effluents with Rhizopus oryzae.

Enzyme preparations are effective in curbing bacterial growth in mill process water and inhibiting the build-up of biofilms. The idea that microorganisms and their enzymes can successfully replace or supplement conventional chemical processes in the pulp and paper industry is gaining acceptance in the United States, following the lead of Scandinavia and Canada. Enzyme preparations have moved from laboratory exploration to mill settings for various industrial applications.

Biotechnology providing new and exciting cost-effective alternatives to complement rather than totally replace traditional technologies. Extensive reviews of the use of enzymes for bleaching, deinking, pitch removal and drainage enhancement illustrate the broad interest in and potential of biotechnology. Relining energy can be significantly reduced if wood chips are pretreated with fungal inocula in the biopulping process or with a bacterial inocula to reduce pitch. All of these technologies have advanced to mill-scale trials. Applications that have been successfully transferred to commercial use include xylanases for bleach boosting, cellulases for improved drainage, lipases for pitch removal, and cellulase-hemicellulase mixture for deinking.

Sustainable Forestry

Selection and hybridisation-Develop optimum natural species and hybrids for use in local conditions. Genetic engineering and tree breeding-Modify gene structure and improve genetic selection programmes to increase growth rates, improve insect and disease resistance, improve fiber quality, and enhance enviromnental adaptability. Following are two emerging applications of biotechnology for forest trees.

1. genetically altering trees for improved forest health through a higher tolerance for pest infestation, disease resistance, and severe climatic conditions, as well as for enhanced properties such as accelerated growth and unproved papermaking qualities (i.e., reduced pitch or lignin content), and
2. phytoremediation of contaminated soil, air, and water.

Recent advances in gene mapping techniques permit researchers to identify trees with desired traits (e.g., fast growth, resistance to disease or cold temperatures) that can be used to breed improved species by conventional means. Mapping permits researchers to focus on specific genes and their components on the molecular level. Identification of gene function permits gene manipulation and the introduction of new and desirable traits not available in the breeding population.

Ultimately, such mapping should permit isolation of desired tree genes that could be engineered directly into target tree species. New techniques for identifying gene markers facilitate the location of desired genes useful for tree breeding Once potentially valuable genes are located, they can be cloned and improved strains of the same or other tree species can be created.

Cloning permits replication of genetically engineered trees and enables mass production of embryos of identical trees that contain one or more value-added traits. Embryos are inserted into manufactured seed and the seeds are sown following conventional culture in a nursery. Identical trees are advantageous in ensuring a uniform raw material that is relatively predictable in its requirements for conversion to pulp and paper. Another approach to genetic altering of trees, which utilises "antisense constructs" to inhibit enzyme production needed for lignin synthesis, has been used by Belgian researchers in their work to minimise the lignin content of trees.

Phytoremediation

Trees are already used for wastewater clean-up, for site stabilisation, and as barriers to subsurface flow of contaminated groundwater. Trees are ideal remediators because they are fast-growing perennial plants with extensive root systems and high transpirational rates. Their large biomass is advantageous because it permits higher tolerance for toxic materials and has the capacity for accumulating contaminants. Because plant remediation is done in situ, it has the potential to be substantially less expensive than alternative technologies used for detoxification. The main methods of phytoremediation are

1. decontamination, in which the amount of toxic pollutants in the soil is significantly reduced or eliminated, and
2. stabilisation and containment in which plants and their associated microflora do not remove contaminants but rather alter the soil chemistry and sequester, reduce, or eliminate the environmental risk of the toxin.

Work is underway to screen tree species for their ability to tolerate, take up, translocate, sequester, and degrade organic compounds and heavy metal ions. Clonal propagation and genetic engineering techniques already exist for a number of species, which opens the door to the creation of tree "remediation" cultivars. This in situ use of plants to stabilise, remediate, and restore a comaminated site is referred to as phytoremediation. All plants have the ability to accumulate metals essential for their growth and development; these metals include iron, manganese, zinc, copper, magnesium, molybdenum, and possibly nickel.

Certain plants accumulate heavy metals that have no known biological function: these metals include cadmium, chromium, lead, cobalt, silver, selenium, and mercury. However, significant accumulation of heavy metals is usually toxic to most plants. For some time, botanists have been aware that certain tree species are endemic to soils containing high metal content. Identification of heavy metal tolerance by some plants has led to research that exploits this characteristic for removing metal contaminants by establishing selected vegetation on contaminated soil; these plants are called "accumulating" plants. A specific example of this technology is the development of a transgenic yellow poplar developed for remediating mercury-contaminated soil.

Hyperaccumulating plants promise effective, inexpensive remediation of soil, sediment, and groundwater. Whereas metal-tolerant plants exclude toxic metal ions from uptake, hyperaccumulating plants take up high amounts of toxic metals and other ions. An exciting possibility of applying biotechnology lies in identifying a tree species with the ability to tolerate or accumulate toxic substances such as heavy metals or organic compounds. Once identified, this tree species could be introduced in contaminated areas. Furthermore, genetic modification could accelerate remediation by making the tree a hyperaccumulator, by adapting its growth to diverse climatic conditions, or by enabling faster growth.

Phytoremediation is based on root uptake of contaminants and storage in the plant or partial/complete degradation to less toxic compounds. This type of remediation could promote degradation of organic pollutants by increasing soil organic carbon content or by releasing enzymes that promote microbial activity through the plant roots. Phytoremediation could be useful in ameliorating heavy metals and organic compounds such as 2,4,6-trinitrotoluene (TNT), trichloroethylene (TCE), benzene, toluene, xylene, and ethylbenzene.

The benefits of phytoremediation include the fact that it is done in situ and that it is a passive, solar-driven "green" technology. Roots are exploratory, liquid-phase extractors that can find, alter, and/or translocate elements and compounds against large chemical gradients. This technology is most effective on sites containing a low level of contamination that are widely dispersed over a large area in the upper surface of the soil. Phytoremediation can work side by side with site restoration with minimum site disruption. Additionally, plant biomass can be harvested to remove contaminants From the site and trees will resprout without disturbing the site. In sites where a valuable heavy metal has accumulated, it may be possible to reclaim the metal from the harvested tree.

Phytoremediation techniques are less expensive than ex situ methods, but they require a long time to work. Longterm site remediation and stabilisation using trees makes remediation and restoration synonymous, which lowers costs and is compatible with public objectives. The most appropriate type of remediation for a specific site depends on the degree of pollution and the type of toxic material. More intensive remedies are required for localised, highly contaminated sites. Conventional soil remediation methods are more suitable for these sites. These methods typically involve excavation of contaminated soil followed by extraction of the toxin. This ex situ technique is usually extremely expensive.

Toxic metal contamination of soil and groundwater is a major environmental and human health problem for which affordable, effective solutions are urgently needed. In agricultural areas, sites are fequently contaminated by a build-up of residual herbicides Atrazine, a commonly used agricultural herbicide, has been the focus of bioremediation researchers. Research with hybrid poplars has resulted in somaclonal variants that tolerate lethal dosages of herbicides. Another aspect of engineered tolerance, pesticide and herbicide resistance, is especially interesting.

If trees could be engineered to be more tolerant of the ubiquitous chemicals in soils, substantially higher yields of forest trees could be realised. Other possibilities for phytoremediation range from removing concentrations of naturally occurring selenium solubilised in irrigation water and accumulated in surrounding groundwater to using genetically altered eucalyptus trees for absorbing and metabolising air pollutants. The possibilities seem to be limited only by the imagination of researchers and the toxic material present.

Applications of Biotechnology in Energy Sector

With regard to the energy sector, one area of considerable importance is the enhanced production of ethanol, either as industrial alcohol or in the brewing and food or beverage industries. The consumption of ethanol in the United States is expected to rise sharply because of the curtailment in the use of lead in gasoline, since ethanol can increase the octane rating by several points as a 10% blend with gasoline. The use of such gasohol has been encouraged for several years, and may assume some importance in the future, depending upon the price of crude oil and its availability.

About 80% of the fuel-grade alcohol in the U.S. is obtained by fermentation with yeasts, and considerable enhancement of such fermentation can be accomplished by genetic engineering. For example, yeasts are poor starch digesters, and in the fermentative production of ethanol from starch, enzymes such as α-amylase and glucoamylase are routinely added as pretreatment of the starch to convert it to linear oligomers and ultimately to glucose units, that are then fermented by the yeasts.

Such fermentative production of ethanol can be considerably enhanced at a cheaper rate if the genes for such enzymes can be cloned in yeast so that the enzymatic pretreatment steps can be avoided. Toward this end, yeast strains containing Aspergillus glucoamylase genes have been constructed where the enzyme is not only glycosylated but is also excreted into the medium. Further improvements of this strain may allow economical fermentation of ethanol directly from starch by genetically engineered yeasts. Another avenue for enhanced production of ethanol by yeasts would be to use cheaper substances such as whey, a by-product of cheese making, or agricultural biomass such as cellulose or hemicellulose.

Considerable progress in the enhanced degradation of pentose sugars from hemicellulose has been reported; while the cellulase genes have been cloned from a number of sources, the development of an efficient cellulolytic yeast is still several years away because of the involvement of a number of enzymes (both endo and exo-glucanases and β-glucosidase) in cellulose degradation. On the other hand, lactose degrading yeast Saccharomyces cerevisiae has recently been constructed by cloning into non-lactose fermenting S, cerevisiae strains the structional gene for β-galactosidase as well as the lactose permease gene. Thus future possibilities for providing ethanol from bulk waste materials look promising.

In addition to the quantity of fuel, the quality of fuels can also be improved by genetic engineering and biotechnology. For example, high sulfur coal and oil, when burned for power generation, often lead to large scale environmental pollution in the form of acid rain. Biotechnological approaches to reducing the concentrations of organic and inorganic sulfur from high sulfur coal and oil have received some attention recently, although large scale processes for the desulfurisation of coal or oil remain a long-term goal.

Microbial enhancement of oil recovery, similarly, has been a subject of much discussion, although formidable problems remain to make a microbial process for oil recovery a reality. A major cause of environmental pollution now-a-days is the release of highly toxic chemical compounds in the form of herbicides and pesticides, solvents, refrigerants and industrially useful compounds such as PCBs (polychlorinated biphenyls) in the environment. Many of these compounds are highly chlorinated, and are extremely persistent in nature.

The persistence of highly chlorinated, synthetic compounds is believed to be due to a lack of utilisation of such compounds by natural microorganisms, where the primary route of metabolism is bacterial co-oxidation, a slow non-energy yielding process. Recently, however, considerable efforts have been directed towards developing microbial strains either by genetic selection or by gene cloning that allows degradation of a number of highly chlorinated compounds. Genetic studies with microorganisms capable of degrading simple chlorinated compounds have demonstrated that the degradative genes are often located on plasmids.

Cloning of the degradative genes for a synthetic compound such as chlorocatechol and the nucleotide sequence homology of the degradative genes with those involved in the degradation of natural analogous compounds such as catechol have demonstrated considerable sequence homology among these genes, suggesting that new degradative functions against synthetic chlorinated compounds may have evolved by recruitment of genes encoding degradation of analogous compounds and then introducing mutational divergence to alter the substrate specificity of the gene products.

The concept that genes encoding the degradation of synthetic chlorinated compounds evolve by recruitment and mutational divergence of genes encoding degradation of their natural structurally-analogous counterparts has allowed us to develop a strain of Pseudomonas cepacia under strong selection in presence of plasmid gene pools in a continuous culture, that can utilise a persistent compound such as 2,4,5-Trichlorophenoxyacetic acid (2,4,5-

T) as its sole source of carbon and energy at a rapid rate. This strain can not only completely dechlorinate 2,4,5-T and its metabolic intermediate 2,4,5-trichlorophenol, but a number of other chlorophenols including pentachlorophenol.

The strain is also very effective in removing large quantities of 2,4,5-T from contaminated soil within a few weeks, suggesting that such strains may be of practical value in removing the toxic chemicals from contaminated sites. An interesting feature of the 2,4,5-T degrading strain is its ability to form surface active compound (s) that would emulsify 2,4,5-T. It is likely that during recruitment and evolution of the 2,4,5-T degradative genes, the genes for the formation of the emulsifier were also recruited to facilitate the uptake of this compound. Normally, highly chlorinated compounds such as PCBs, TCDD (2,3,7,8-tetrachloro-dibenzo-p-dioxin) etc. are very hydrophobic, and are unavailable to the bacterial cells because of their insolubility in water.

The hydrophobic nature of most highly chlorinated compounds must have contributed to the persistence of the compounds, since microorganisms need not only to evolve the degradative genes for these compounds, but also genes for the synthesis of appropriate emulsifiers to facilitate their uptake in an aqueous environment. This situation is reminescent of the microbial degradation of hydrocarbons present in crude oil, most of which are also highly hydrophobic.

11

Impact of Biotechnology on Biodiversity

Biodiversity today is the result of 3.5 billions years of evolution. All the living organisms we know today, as well as those that ever lived before, have developed from one original micro-organism through the processes of mutation and selection. Separate species arose when mutations between relatives no longer allowed for interbreeding, for instance after geographic separation. The vast majority of species that arose, probably more than 99%, disappeared again. In the long-term view there has been no such thing as sustainability, only change. Clearly however today, with man's massive influence on the globe, change and the loss of biodiversity is much faster than at any time before, making concern with sustainability important.

Biodiversity can be distinguished at three different levels, ecosystems, species and genes. But there is no generally accepted definition of biodiversity and there is no general consensus on how to appreciate changes in biodiversity from scientific, political and/or normative perspectives. The number of species of animals, plants and micro-organisms today is probably 10 million or more, of which only 1.4 million have yet been named. Virtually all the 40,000 vertebrate animals and most of the 250,000 higher plant species are known.

On the other hand there are likely to be over a million species each of fungi and nematodes (thread-like worms), and several million insects, of which only 70,000, 13,000 and 950,000 respectively have been identified. Only about 5,000 bacteria and viruses have been identified individually, yet their total number may also be well in excess of one million. The various species of plants and animals do not live an independent existence but are associated in specific communities and ecosystems to form more or less stable associations.

One such association, for instance, is the tropical rain forest which is generally thought to have the highest degree of biodiversity. Biodiversity needs to be considered both in terms of the number of species present and also of how many individuals there are of each of the species. Often the number of species in a given ecosystem is taken as a measure of the biodiversity of that system because other criteria are more difficult to apply.

In addition, it is still not clear whether the total amount of living matter at a given location, the biomass, is generally dependent on biodiversity: some researchers claim that loss of species is not compensated by additional growth of other species and thereby leads to a reduction in total biomass. In agriculture, some 7,000 species of plants are used by farmers throughout the world, but only 30 species provide 90% of our calorific intake, the top three crops being wheat, rice and maize (corn).

Within the predominant crop species there are many hundreds of thousands of varieties (landraces) adapted to local climates and farming practices. Much of this large crop diversity is important for providing starting material for breeding. However, the genetic diversity in crops is much less broad than that of plants or animals living in the wild which indicates the importance of wild species for agricultural breeding programmes.

Growth of Population

The world population has risen from 2.5 billions in 1950 to 6 billion today and is expected to reach some 9 billion in 2050. 800 million people are currently undernourished. Over 95% of the expected population increase will be in the less developed countries (LDCs) and most will occur in their cities, requiring more living space, water, energy, wood, food and services. For 1995 to 2020 the largest relative population increase (80%) is expected in sub-Saharan Africa rising from 500 to 900 million. Whether the HIV/AIDS epidemic in this region will substantially affect population dynamics is not clear as most of those currently infected are in the reproductive age range.

Loss of Biodiversity

Loss of biodiversity can be measured by a loss of individual species, groups of species or by decreases in numbers of organisms. The major threats to global biodiversity are, firstly, habitat loss (mostly through expansion of cultivated land and of cities and roads) and, secondly, the introduction of exotic species. Habitats can also be damaged by flooding, lack of water, climate changes etc., natural or man-made. Since tropical humid forests are particularly rich in biodiversity, their destruction is disproportionately damaging. It is estimated that only half of the original 16 million km^2 of these forests a century ago are left, with about one million km^2 being destroyed every 5 to 10 years.

Biodiversity is not homogeneously distributed but rather there are areas with particularly plentiful biodiversity ("hotspots") of particular interest for conservation. Imported plant and animal species threaten the native ones by being highly competitive and often by lacking local predators. One of the most extreme examples is seen in the pampas of Argentine, a flat grassland with a moderate climate, from which nearly all the native grasses have disappeared and have been replaced by European plants. This invasion was brought about by European farmers, bringing animals and crops, as well as accidentally spreading many different weeds and was already noted in 1833 by Charles Darwin.

Islands are particularly threatened by invaders, as is well documented for Hawaii, New Zealand or the Galapagos Islands. Biological control agents are often introduced into agricultural ecosystems intentionally to control pests or weeds without resorting to chemical controls. Whilst there have been many welcome successes, such systems may also go wrong. One example is the introduction of the seven-spot ladybird which was intended to control the Russian wheat aphid in the US. This proved to be a competitor of the native ladybird, which then disappeared.

The mongoose, an Indian mammal, was introduced to several islands (Fiji, Mauritius, Hawaii) to keep rats and snakes under control: it lead to the extinction of several endemic birds, reptiles and amphibians. Introduced wasps have been seen to kill and possibly extinguish endemic butterflies. Are transgenic plants as such prone to spread? In the longest term experiment so far, four different crops (oilseed rape, potato, maize and sugar beet) were grown in 12 different habitats and monitored over a period of 10 years.

In no single case were the genetically modified plants found to be more invasive or more persistent than their conventional counterparts. However, one would not expect this to be the case unless the transgene increased its fitness in the wild. There is no plausible reason why crops that have for centuries depended for survival on human care should become weeds because of the addition of one or a small number of well-characterised genes, in addition to the many thousands of genes they already carry. Even so, monitoring of transgenic crops needs to continue for more than ten years. Large acreages of a single variety without rotation, should be avoided with any crop, since monocultures are more prone to disease and pest outbreaks.

STRATEGIES FOR CONSERVATION

Conservation may be in situ in a more or less natural or habitat or ex situ in some purposebuilt environment, depending on the particular case. In situ conservation involves the maintenance and protection of natural habitats while botanical gardens, seed banks and zoos are used for ex situ conservation. Conserving a substantial fraction of the tropical rain forests would still allow half or so of their indigenous species to be preserved by selection of the most appropriate "hot spots". Protecting large tracts of land poses major socio-economic and political problems.

How forests can be kept free from encroachment by hungry people in search of potential farmland is far from clear. A viable strategy may be to find sustainable livelihoods for rural populations compatible with conserving tropical rain forests and this is being attempted with mixed success. Policing alone will not be successful over vast territories, as seen in the combating of drugs. Conservation also embraces agricultural biodiversity such as crop varieties, land races (local varieties), semi-domesticated varieties and crop relatives. The role of indigenous communities in maintaining agrobiodiversity is stressed by the Global Biodiversity Assessment and the Leipzig Plan of Action, two recently concluded international agreements.

THE CONVENTION ON BIOLOGICAL DIVERSITY (CBD)

The United Nations adopted the Convention in 1992 and it entered into force in 1993. This is the first time that a large majority of States, though not the USA, have agreed to a legally binding instrument for biodiversity conservation and the sustainable use of biological resources. A radical change brought about by the CBD is the recognition that States have a sovereign right over biodiversity within their own territory, while previously living organisms were considered the common heritage of mankind. Under the terms of the CBD, living organisms or their products may only be removed from a country under mutually agreed conditions.

Action is delegated to the national level obliging States to assess biodiversity, enact legislation for its conservation in situ and ex situ, and to enforce legislation within national boundaries. Biotechnology is particularly affected by Articles 16 and 19 of the CBD since they require a fair and equitable sharing of benefits derived from the use of genetic resources. This includes providing facilities and financial means for technology transfer and open access to scientific and technical information.

The CBD has only been in operation for a few years and so it is too early to assess its long-term effects. As far as biotechnology and "bioprospecting" are concerned, it will take time to establish smooth administrative procedures to allow simple routine implementation of close collaborations. Only if national authorities from countries rich in biodiversity as well as pharmaceutical and other companies see the advantages to be gained from such collaboration will the system expand. One example of a joint effort in "bioprospecting" is the search for specific active ingredients of plants by the Merck Company in the tropical forests of Costa Rica which brought the country 2 million USA dollars over a five year period as well as the potential of royalties if profitable products emerge.

The Cartagena Protocol on Biosafety The CBD provided a basis for developing an international agreement regulating the crossboundary movement of living GMOs, called "living modified organisms" in the Protocol, which after a lengthy debate was agreed on in Montreal in early 2000. The main aim of the Protocol is to ensure that GMOs would not endanger biodiversity in the recipient countries; in addition it also covers potential threats to human health. In practice the Protocol will primarily affect the importation of transgenic seeds. It requires the exporting country to provide enough scientific information to allow the importing country to judge risks and then to issue, if the product is judged safe, an "Advance Informed Agreement".

A newly established Biosafety Clearing House collects and distributes relevant information, as defined by the Protocol. Based on the "Precautionary Principle" import licences can be refused, even when there is no clear-cut proof that the product is dangerous to human health or the environment, but such decisions need to be based on a risk assessment. The Protocol is construed to be as important as the World Trade Organisation (WTO) agreements which do not allow countries to prevent the import of products unless there are sound scientific reasons showing that the product may cause harm to human health. This apparent lack of priority between the CBD and WTO agreements may lead to some future difficulties. The Protocol will become operational when it is ratified by 50 nations, presumably in 2002.

Biotechnology can be used as a tool for acquiring scientific knowledge or to intervene directly in plant and animal breeding and, in particular, to transfer genetic information from any other organism to a crop or farm animal. Taxonomy, one of the sciences necessary to study biodiversity, uses molecular markers to identify individual strains of organisms or to identify species in much the same way as is done in forensic medicine to identify criminals. In seed banks genetic fingerprints are used to establish the origin of a seed or the relatedness of plant varieties.

Biotechnology is also useful for following genetic markers in plant and animal breeding done by conventional means. By analysing a few cells of the newly born calf or of the newly sprouted crop and looking for the presence or absence of certain genes it is possible to predict properties of the progeny which will show up only later in life such as the characteristics of a cow's milk or a crop's expected resistance to a plant disease. These applications of biotechnology to farm animals (not to humans) are hardly controversial.

In 2000 there were about 44 million hectares worldwide planted with commercial transgenic crops mostly in the USA, Canada, Argentina, with smaller amounts in China, Australia, South Africa, Mexico and Spain. Soybean and corn ranked first and second with 25.8 and 10.3 million hectares. Cotton and canola accounted for about 5 and 3 million hectares, whilst only small areas of transgenic potato, squash and papaya were grown commercially. Herbicide tolerant varieties dominated with 74%, while insect resistant varieties made up 19%.

Virus resistant crops were quite small. USA farmers adopted transgenic crops rapidly because of the economic benefits they offered. The most noticeable difference to the farmers was the saving on herbicides. Many also found the new crops needed less frequent sprayings and allowed "no till" (no ploughing and no harrowing) management.

These benefits may offset the initial higher costs of the new transgenic seeds. Clearly these considerations only hold for countries with agricultural and economic systems similar to those of the USA and not for developing countries.

Whether in the near future there will be an expansion or a reduction in the area of transgenic crops grown will depend on market forces, the public perception of risks and benefits, as well as on the emerging national and international regulatory framework. A large number of transgenic crops are in the development stage, including many tropical crop plants; most of them will only come on the market in a few years, if at all. They are likely to show benefits for the consumers and some may be of particular interest to farmers in tropical countries. Two rice varieties currently being developed should lead to important health benefits.

"Golden Rice" contains increased levels of vitamin A and another variety has more iron in the grains than usual, a clear benefit to anaemic women and their children. Several lines of transgenic farm animals have been produced but none have been commercialised. Some lines are made for the pharmaceutical industry to produce drugs in their milk. Others may show improved resistances towards certain infections or produce novel enzymes. Transgenic salmon that grow faster than normal have been developed and have roused considerable concern amongst ecologists, particularly regarding their potential to compete with native salmon in the wild.

Here many environmental issues still need clarification. Native biodiversity and biotechnology Biodiversity in the wild has been massively reduced in the industrialised countries over many centuries and about half the tropical rain forests have already been destroyed.

How then can the rest be preserved, given that the world population is increasing rapidly and more food has to be produced regionally? Yields of cereals in LDCs have gone up very considerably in the last forty years primarily as the result of the Green Revolution.

However, the annual growth increases in cereal yields in LDCs have slowed down from about 3% during 1967-82 to about 1% per year from 1993 onwards. This lowering of the rate at which yields have increased means that productivity will probably not keep up with the demands of increased populations. The consequences for biodiversity are devastating as more land will be required for farming which will primarily come from areas with high native biodiversity, and in particular tropical rain forests.

The single most promising way to avoid habitat destruction is to increase farm yields in a process that has been called the "Second Green Revolution". Several components will be required including training and education of farmers (in particular of women who do most of the farm work in LDCs), more favourable economic and political climates, availability of farm credit schemes, etc. In addition, technical contributions will be necessary and, in particular, improved seed produced either by traditional crop breeding or by modern biotechnology. Reliance will have to be more on the latter since traditional breeding appears to have reached a plateau in yield and is slower, less precise and only feasible when interbreeding is possible.

So agricultural biotechnology, which is viewed frequently in the public debate as harmful to biodiversity, is, paradoxically, likely to contribute to conserving it. A more limited concern that largely affects Northern Europe is the conservation of native plants and animals, in particular birds, in farmed areas. Their habitats are fields, hedges, roadsides and fallow land where they depend for food on insects and seeds produced by weeds in or near crops.

Computer models suggest that more intense weed control measures may lead to smaller amounts of seeds being available to birds. This depends far more on weed management regimes rather than on transgenic plants. Herbicide tolerant beets allow farmers to control weeds later by treating after the seedlings have emerged. The more efficient methods may allow setting aside of more land. Setting aside more farmland requires financial incentives.

Biotechnology and Agricultural Biodiversity

Gene transfer between cultivated and wild plants has always occurred within the limits of species if the two types of plants were in close proximity and flowered at the same time. No new problems can be expected from transgenic plants except if the gene transferred from the GMO to the wild plant significantly increased the biological fitness of the recipient. This generally seems unlikely a priori and can, and should, be checked experimentally by risk assessment.

Some people feel there is a crucial difference between transfer of genes from conventionally bred plants to their wild relatives and transfer of man-made transgene constructs from a GM plant into a gene pool. Whether ecological risks may result from the introduction of a man-made gene construct into the gene pool of a species is a different question. Farmers all over the world use a remarkable number of landraces and varieties of many different crops. These are usually closely adapted to the local climate and topography, and are also used to produce foods for different purposes.

To judge by what has happened on the US soybean market, this will not be the case. Although virtually all herbicide resistant soybeans derive from a single transformation event in Monsanto's Roundup Ready soya, hundreds of different varieties have been derived from it by different seed companies using traditional breeding to develop soya varieties that suit different climatic and soil regions. This shows that at least with this dominant transgenic crop biotechnology has not led to a loss of agricultural biodiversity. On the other hand, judging from the last 100 years in Europe, the diversity of varieties of farm animals and crops has, in fact, diminished considerably, but not because of any biological hazard emanating from some of the breeds, but because farmers need to produce economically.

Many old varieties of apples have disappeared, for example, because of the preference of food retailers and consumers. The rapid consolidation in the global seed market may be a concern in this respect and anti-trust legislation may be needed to prevent the dominance of oligopolies with limited competition between few producers. It can be argued that only wealthy farmers will be able to afford transgenic seeds. However, the technology is well suited to small farmers, since it is packaged in a small seed grain, which can be reproduced locally. However, if the seeds are to reach small farmers, there need to be many seed outlets (public cooperatives or private companies) and adequate agricultural extension programmes.

Under what economic and political conditions will biotechnology benefit agricultural productivity and thereby help biodiversity? The original green revolution which improved wheat and rice production in Asia was initiated by research carried out in the public sector. Today's innovative research in biotechnology is mainly done by a few large companies with modest contributions from public sector and small companies. Companies strive for. intellectual property rights (IPR) through patenting for return on their research investment. This is largely unavoidable with the present IPR systems.

Provisions need to be made so that the agricultural research institutions of LDCs obtain access to information as well as patented materials and procedures needed for regional farmers. Unfortunately financial support to the agricultural research systems of LDCs has been substantially cut in the last ten years: this trend endangers world food security and needs to be reversed. A limited number of transfers of specific patents to LDCs has already occurred from companies such as Monsanto and Syngenta for producing virus-resistant sweet potatoes and Vitamin A enriched golden rice.

This approach; including the possibility of enforced licensing, is probably more practicable than changing the world's patenting system which does, however, need to be adapted in order to deal with living organisms. A new issue arising in the field of IPR is the attempt to protect traditional, indigenous knowledge through novel international legislation. Lack of food appears to be, and is largely, a problem of distribution; there is not sufficient food available to those who need it.

The poor simply do not have the money to buy enough. Since in LDCs most of the poor still live on the land, it is here that production has to be increased so that there is not only enough for survival but a surplus which can be sold to buy other goods and services. Increasing local agricultural productivity is the most effective way to do this and should, according to many economists of the public sector including those of the UNDP, embrace modern biotechnology as an important component.

Bibliography

Babu P.S and Panda T. "Studies on improvement techniques for immobilizing and stabilizing penicillin amidase associated with *E. coli* cells", *Enzyme Microbial Technology*. 13: 676-682, 1991.

Baianu I. C. "Computer models and automata theory in biology and medicine". *Mathematical Modelling*. 7: 1513-1577, 1986.

Baianu I.C, Magin R .and Ozu E.. "Microencapsulation adds value to Illinois crops", *Illinois Research*. 35(1):19-21, 1993.

Chawla, H.S., *Introduction to Plant Biotechnology* Oxford and IBH 2000.

Dastagiri M.B. "Demand and policy projections for livestock production in India". *Policy paper* No. 21. National Centre for Agricultural Economics and Policy Research, New Delhi. 2004.

Delgado C.L., Hopkins J. & Kelly V.A. "Agricultural growth linkages in sub-Saharan Africa. Research Report" No. 107, *International Food Policy Research Institute*, Washington, DC, 154 pp. 1998

Denning C. & Priddle H. "New frontiers in gene targeting and cloning: success, application and challenges in domestic animals and human embryonic stem cells", *Reproduction*, 126, 1-11, 2003.

Dhalmini Z., Spillane C., Moss J.P., Ruane J., Urquia N. & Sonnino A. "Status of research and application of crop biotechnologies in developing countries – a preliminary assessment", Renouf Books, Ogdensburg, New York, 62 pp. 2005.

Dorjee K., Broca S. & Pingali P. "Diversification in South Asian agriculture: trends and constraints". *ESA Working paper* 03-15. Food and Agriculture Organization, Rome, 23 pp. 2003.

Drucker A.G. "The economics of farm animal genetic resource conservation and sustainable use: why is it important and what have we learned?" *Background study paper* No. 21. Food and Agriculture Organization, Rome, 11 pp. 2004.

Dubey, R.C, S.Chand & Co.Ltd. *A Text book of Biotechnology*, 1993.

Sang O.Y, Twaiq F, Zakaria R, Mohamed A R, Bhatia S. "Biofuel production from catalytic cracking of palm oil". *Energy Sources* 25:859-869, 2003.

Sheehan J, Himmel M.E "Enzymes, Energy, and the Environment: Cellulase Development in the Emerging Bioethanol Industry". *Biotechnol. Progress* 15:817-827, 1999.

Solomon G, Ogunseitan O.A, Kirsch J. "Pesticides and Human Health: A Resource for Health Care Professionals". Physicians for Social Responsibility, Los Angeles and Californians for Pesticide Reform, San Francisco, 2000.

Sullia S.B., and S.Shantaram *General Microbiology*, Oxford and IBH 1998.

Thomas V, Theis T, Lifset R, Grasso D, Kim B, Koshland C, Pfahl R. "Industrial Ecology: Policy Potential and Research Needs". Environ. Engr. Sci. 20:1-9, 2003.

Thomas V.M, Kwong A. "Ethanol as a Lead Replacement: Phasing Out Leaded Gasoline in Africa". *Energy Policy,* 29:1133-1143, 2001.

Index